T0275741

Digital Holographic Data Representation and Compression

Digital Holographic Data Representation and Compression

Yafei Xing

Mounir Kaaniche

Béatrice Pesquet-Popescu

Frédéric Dufaux

AMSTERDAM • BOSTON • HEIDELBERG • LONDON
NEW YORK • OXFORD • PARIS • SAN DIEGO
SAN FRANCISCO • SINGAPORE • SYDNEY • TOKYO
Academic Press is an imprint of Elsevier

Academic Press is an imprint of Elsevier
125 London Wall, London, EC2Y 5AS, UK
525 B Street, Suite 1800, San Diego, CA 92101-4495, USA
225 Wyman Street, Waltham, MA 02451, USA
The Boulevard, Langford Lane, Kidlington, Oxford OX5 1GB, UK

Notices
Knowledge and best practice in this field are constantly changing. As new research and experience broaden our understanding, changes in research methods, professional practices, or medical treatment may become necessary.

Practitioners and researchers must always rely on their own experience and knowledge in evaluating and using any information, methods, compounds, or experiments described herein. In using such information or methods they should be mindful of their own safety and the safety of others, including parties for whom they have a professional responsibility.

To the fullest extent of the law, neither the Publisher nor the authors, contributors, or editors, assume any liability for any injury and/or damage to persons or property as a matter of products liability, negligence or otherwise, or from any use or operation of any methods, products, instructions, or ideas contained in the material herein.

Library of Congress Cataloging-in-Publication Data
A catalog record for this book is available from the Library of Congress

British Library Cataloguing in Publication Data
A catalogue record for this book is available from the British Library

For information on all Academic Press publications visit our website at http://store.elsevier.com/

ISBN: 978-0-12-802854-4

CONTENTS

CONTENTS

LIST OF TABLES

LIST OF FIGURES

ACRONYMS

1D	one dimensional
2D	two dimensional
3D	three dimensional
ASP	advanced simple profile
ASQ	adaptive scalar quantization
AVC	advanced video coding
BS	beam splitter
CCD	charged coupled device
CGH	computer-generated holography
CPU	central processing unit
DA-DWT	direction-adaptive discrete wavelet transform
DCT	discrete cosine transform
DFT	discrete Fourier transform
DH	digital holography
DWT	discrete wavelet transform
EBCOT	embedded block coding with optimal truncation
EZW	embedded zerotree wavelet
GPU	graphics processing units
HEVC	high efficiency video coding
LBG	Linde-Buro-Gray
LIP	list of insignificant pixels
LIS	list of insignificant sets

LS	lifting scheme
LSP	list of significant pixels
MDC	multiple description coding
MSE	mean squared error
NRMS	normalized root mean square
NS-VLS	nonseparable vector lifting scheme
PSDH	phase-shifting digital holography
PSI	phase-shifting interferometry
PSNR	peak signal-to-noise ratio
PZT	piezoelectric transducer
SPIHT	set partitioning in hierarchical trees
SQ	scalar quantization
SSIM	structural similarity
USQ	uniform scalar quantization
VLC	variable length codewords
VLS	vector lifting scheme
VQ	vector quantization

CHAPTER 1

Introduction

Thanks to tremendous technological advances over the last decades, imaging systems are rapidly evolving, with each generation delivering greatly enhanced visual quality. In this context, three-dimensional (3D) imaging is often considered as a key feature in order to provide a more lifelike and immersive user experience (Dufaux et al., 2013).

However, most existing 3D technologies rely on stereoscopic or multiview representations. Unfortunately, these approaches have some fundamental limitations and only provide a limited set of depth-cues. As a consequence, user experience often remains below expectations. In addition, the inherent accommodation-vergence conflict, that is, the eyes are focusing on one point but converging toward another one, may induce headaches, nausea, or visual fatigue. These significant drawbacks explain the difficulty of 3D imaging technologies to gain momentum and broad market adoption in some application domains.

Holography was invented by Dennis Gabor in 1948 (Gabor, 1948). A hologram records the light field emanating from an object as an interference pattern. It subsequently allows reproducing this same light field and therefore reconstructing 3D views of this object which are theoretically indistinguishable from the original ones. Therefore, holography has the potential to be the ultimate 3D experience, with continuous head motion parallax, natural eyes accommodation and vergence, and all depth-cues including perspective, occlusions, lighting, shading, and defocus blur.

It is also worth mentioning two major milestones in the development of holography. The emergence of digital technologies made digital holography (DH) (Goodman and Lawrence, 1967; Schnars and Jüptner, 1994) possible. More specifically, rather than using a photographic recording, with DH the hologram is numerically recorded by a light-sensitive sensor such as a charged coupled device. Next, with the continuous computing power increases of modern computers, it became possible to simulate computationally the whole procedure. This is referred to as computer-generated holography (CGH) (Dallas, 1980; Tricoles, 1987).

Digital Holographic Data Representation and Compression. http://dx.doi.org/10.1016/B978-0-12-802854-4.00001-X

Holography has been successfully applied in a number of applications, notably interferometric microscopy and metrology. However, major scientific and technological challenges remain before holography can fulfill its potential to become the ultimate 3D experience. For instance, signal processing issues for holographic 3DTV are discussed in Onural et al. (2007) and Onural and Ozaktas (2007). The development of holographic displays is also a key aspect for successful applications. Recent developments are presented in Bove (2012).

In this book, we more specifically consider the problem of representation and compression of holographic data. Indeed, digital holographic data represent a huge amount of information which has to be efficiently handled. In addition, a very high data rate is needed for a real-time system. These two drawbacks constitute major bottlenecks. Clearly, efficient compression of holographic visual information is a critical aspect. Given that the signal properties of holograms significantly differ from those of natural images and video sequences, merely applying existing compression techniques (e.g., JPEG or MPEG standards) remain suboptimal. This field of research is still relatively new and clearly calls for innovative compression solutions. This book aims at presenting a comprehensive overview of state-of-the-art compression techniques for digital holographic data, along with a critical analysis.

The book is structured as follows. In Chapter 2, fundamental principles of DH are reviewed. In particular, we present physical principles, CGH, and phase-shifting DH. Basic compression tools are briefly discussed in Chapter 3. Chapter 4 introduces and compares different representations of digital holographic data. The compression of digital holographic data is addressed in Chapter 5. More specifically, we review some quantization-based and wavelet-based approaches. Finally, some concluding remarks as well as open issues are discussed in Chapter 6.

CHAPTER 2

Fundamental Principles of Digital Holography

In order to have a better understanding of holography, this chapter outlines the main aspects relevant for the optical and numerical imaging principles of DH. Two main physical principles—interference and diffraction—of holography are first introduced which correspond to the procedures of hologram acquisition and its reconstruction. Based on the physical principles, computational methods for generating digital holograms and numerical reconstruction have been developed.

2.1 PHYSICAL PRINCIPLES

Figure 2.1 shows a typical in-line geometry for recording holograms. Several beam splitters, mirrors, and lens are used to manipulate the light wave. A laser beam is first split into two paths by a beam splitter. One beam illuminates the object forming a scattered object wave along the axis normal to the photographic plate (hologram plane). The other one, which is called the reference wave, is led to be parallel to the object wave and interfere with the object wave at the hologram plane. The recorded hologram is obtained from the interference pattern.

Digital Holographic Data Representation and Compression. http://dx.doi.org/10.1016/B978-0-12-802854-4.00002-1

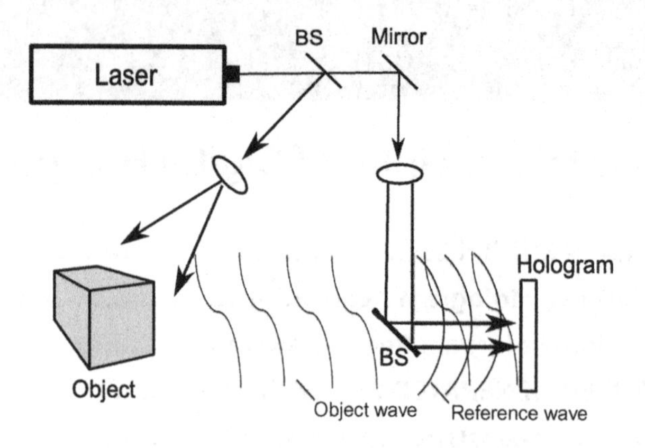

Figure 2.1 Hologram recording setup. BS, beam splitter.

2.1.1 Interference-Hologram Acquisition

In accordance with Maxwell equations, the complex amplitude of the object wave U_O and the reference wave U_R at the hologram plane can be represented by

$$\begin{aligned} U_\mathrm{O}(x,y) &= A_\mathrm{O}(x,y)\exp[\mathrm{i}\varphi_\mathrm{O}(x,y)], \\ U_\mathrm{R}(x,y) &= A_\mathrm{R}(x,y)\exp[\mathrm{i}\varphi_\mathrm{R}(x,y)], \end{aligned} \tag{2.1}$$

respectively, where (x,y) is the coordinate of one point at the hologram plane, and $(A_\mathrm{O}, \varphi_\mathrm{O})$ and $(A_\mathrm{R}, \varphi_\mathrm{R})$ are the amplitude and phase distribution pairs of the object wave and reference wave, respectively. The resultant intensity $I(x,y)$ of the interference pattern is expressed as

$$\begin{aligned} I(x,y) &= |U_\mathrm{R}(x,y) + U_\mathrm{O}(x,y)|^2 \\ &= U_\mathrm{R}(x,y)U_\mathrm{R}^*(x,y) + U_\mathrm{O}(x,y)U_\mathrm{O}^*(x,y) + U_\mathrm{R}(x,y)U_\mathrm{O}^*(x,y) \\ &\quad + U_\mathrm{O}(x,y)U_\mathrm{R}^*(x,y), \end{aligned} \tag{2.2}$$

where $*$ represents the conjugate complex. The amplitude transmission $I_\mathrm{H}(x,y)$ recorded on the hologram plane is

$$I_\mathrm{H}(x,y) = h_0 + \beta\tau I(x,y), \tag{2.3}$$

where h_0 is the amplitude transmission of the unexposed plate, β is a constant, and τ is the exposure time. The amplitude and phase information of the object wave are thus recorded in the hologram.

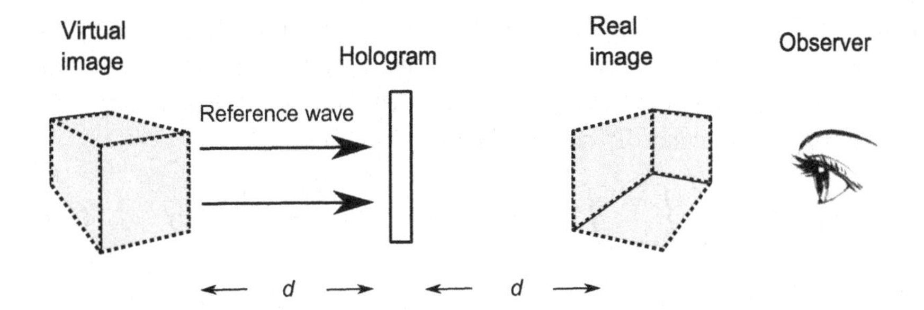

Figure 2.2 Hologram reconstruction.

2.1.2 Diffraction-Hologram Reconstruction

The object image can be optically reconstructed as shown in Fig. 2.2 by illuminating the hologram using the same reference wave. A virtual image can be observed at the same position as the original object while the real image is formed in the opposite direction at a distance d from the hologram. If the hologram is illuminated by the conjugate reference wave, the real and virtual images can be observed in the opposite positions than where they are observed in Fig. 2.2. The coordinate geometry of Fig. 2.2 can be explained in Fig. 2.3.

The reconstruction procedure is considered as a diffraction phenomenon which happens when a light wave (the reference wave) illuminates an aperture (the hologram) in an opaque plane. It can be explained by scalar diffraction theory if two conditions are met (Goodman, 1996): (1) the diffracting aperture must be large compared with the wavelength, and (2)

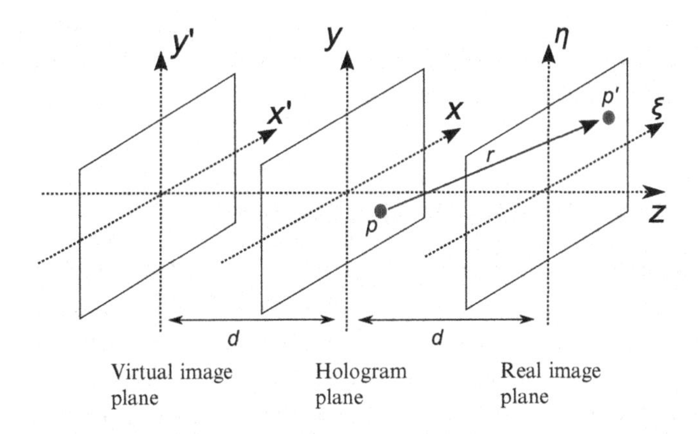

Figure 2.3 Coordinate geometry of hologram reconstruction.

the diffracting fields must not be observed too close to the aperture. Suppose that the reference wave is perpendicular to the hologram plane as shown in Fig. 2.2, the resulting diffraction can be described by the Fresnel-Kirchhoff integral under the scalar diffraction theory (Goodman, 1996):

$$U(\xi,\eta) = -\frac{\mathrm{i}}{\lambda}\int_{-\infty}^{+\infty}\int_{-\infty}^{+\infty} I_{\mathrm{H}}(x,y)U_{\mathrm{R}}(x,y)\frac{\exp(\mathrm{i}kr)}{r}\,\mathrm{d}x\,\mathrm{d}y \tag{2.4}$$

with

$$r = \sqrt{(x-\xi)^2 + (y-\eta)^2 + d^2}, \tag{2.5}$$

where $U(\xi,\eta)$ is the complex amplitude of the real image at the real image plane (ξ,η), λ is the wavelength of the reference wave, $I_{\mathrm{H}}(x,y)$ is the hologram function in the hologram plane $(Z=0)$ at point P with coordinate (x,y), r is the distance between point P in the hologram plane to point P' with coordinate (ξ,η) in the real image plane $(Z=d)$, and d is the distance between the two planes (real image and hologram plane, respectively, virtual image and hologram plane).

Moreover, as the hologram plane can be considered planar, the reconstruction diffraction can also be interpreted by Rayleigh-Sommerfeld diffraction formula (Goodman, 1996).

$$U(\xi,\eta) = -\frac{\mathrm{i}}{\lambda}\int_{-\infty}^{+\infty}\int_{-\infty}^{+\infty} I_{\mathrm{H}}(x,y)U_{\mathrm{R}}(x,y)\frac{\exp(\mathrm{i}kr)}{r^2}\cos\theta\,\mathrm{d}x\,\mathrm{d}y, \tag{2.6}$$

where θ is the angle between the vector $\overrightarrow{PP'}$ and the surface normal to the hologram plane (or the propagation direction along axis z) as shown in Fig. 2.3. Note that even though Kirchhoff theory is more general than the Rayleigh-Sommerfeld one, which is because the latter one requires planar diffraction screen while the former does not, in the case of hologram reconstruction, they are both practical.

2.2 COMPUTER-GENERATED HOLOGRAPHY

As explained before, the general idea of DH includes optical recording and digital processing. However, with the improvement of modeling virtual objects by computer, reconstructing virtual objects or scenes wavefront by holographic representation has become a popular topic. Even though there are still some limitations, such as sampling rates, computation, and storage burden, the future of CGH is very promising. In this section, the main concepts of CGH are addressed. In particular, the general numerical reconstruction methods for DH are first explained.

2.2.1 Numerical Reconstruction Methods

As introduced before, the reconstruction procedure can be described by both the Fresnel-Kirchhoff theory and Rayleigh-Sommerfeld theory. Based on that, two kinds of numerical reconstruction methods will be addressed: the approximation methods based on Fresnel-Kirchhoff theory and the angular spectrum method based on Rayleigh-Sommerfeld theory.

The approximation methods can reduce the diffraction pattern calculations to simpler mathematical manipulations. Based on these approximations, numerical processing can be easily conducted by using Fourier transform.

2.2.1.1 Approximation Methods

Reconstruction by Fresnel approximation The Fresnel approximation is obtained by approximating the distance r introduced in Eq. (2.5). More specifically, the expression can be expanded as a Taylor series:

$$r = d + \frac{(x-\xi)^2}{2d} + \frac{(y-\eta)^2}{2d} - \frac{1}{8}\frac{[(x-\xi)^2 + (y-\eta)^2]^2}{d^3} + \cdots. \quad (2.7)$$

Then, it is approximated to

$$r \approx d + \frac{(x-\xi)^2}{2d} + \frac{(y-\eta)^2}{2d}, \quad (2.8)$$

if the fourth term is very small compared to the wavelength:

$$\frac{1}{8}\frac{[(x-\xi)^2 + (y-\eta)^2]^2}{d^3} \ll \lambda. \quad (2.9)$$

Substituting Eq. (2.8) into Eq. (2.5) with the additional approximation of the denominator by d, the resulting complex amplitude of the real image can be expressed as follows:

$$U(\xi,\eta) = -\frac{\mathrm{i}}{\lambda d}\exp(\mathrm{i}kd)\int_{-\infty}^{+\infty}\int_{-\infty}^{+\infty} I_{\mathrm{H}}(x,y)U_{\mathrm{R}}(x,y) \times \exp\left[\frac{ik}{2d}((x-\xi)^2+(y-\eta)^2)\right]\mathrm{d}x\,\mathrm{d}y. \quad (2.10)$$

Factoring the term $\exp\left[\frac{ik}{2d}(\xi^2+\eta^2)\right]$ outside the integral yields

$$U(\xi,\eta) = -\frac{\mathrm{i}}{\lambda d}\exp\left[\mathrm{i}k\left(d+\frac{\xi^2+\eta^2}{2d}\right)\right] \times \int_{-\infty}^{+\infty}\int_{-\infty}^{+\infty} I_{\mathrm{H}}(x,y)U_{\mathrm{R}}(x,y) \times \exp\left[\frac{ik}{2d}(x^2+y^2)\right]\exp\left[-\frac{ik}{d}(x\xi+y\eta)\right]\mathrm{d}x\,\mathrm{d}y, \quad (2.11)$$

where the mathematical similarity with the Fourier transform can be easily recognized. This formulation is named Fresnel approximation, where the observer is in the near field of the aperture.

Reconstruction by Fraunhofer approximation The stronger Fraunhofer approximation comes in addition to the Fresnel approximation. If the condition

$$d \gg \frac{k(x^2+y^2)}{2} \quad (2.12)$$

is satisfied, then the quadratic phase factor $\exp\left[\frac{ik}{2d}(x^2+y^2)\right]$ under the integral sign in Eq. (2.11) is approximately equal to unity. The reconstructed complex amplitude of the real image is consequently substituted by

$$U(\xi,\eta) = -\frac{\mathrm{i}}{\lambda d}\exp\left[\mathrm{i}k\left(d+\frac{\xi^2+\eta^2}{2d}\right)\right] \times \int_{-\infty}^{+\infty}\int_{-\infty}^{+\infty} I_{\mathrm{H}}(x,y)U_{\mathrm{R}}(x,y)\exp\left[-\frac{ik}{d}(x\xi+y\eta)\right]\mathrm{d}x\,\mathrm{d}y, \quad (2.13)$$

which can also be obtained similarly from the Fourier transform of the aperture distribution $I_{\mathrm{H}}(x,y)U_{\mathrm{R}}(x,y)$. It is named Fraunhofer approximation where the observer is located in the far field of the aperture.

Discrete reconstruction For the sake of simplicity, the derivation of discrete reconstruction is based on a propagation between two parallel square planes (the hologram plane and the real image plane) without offset. The extension to rectangles is straightforward. For digitizing the reconstruction formulas, the following substitutions are introduced:

$$\nu = \frac{\xi}{\lambda d}; \quad \mu = \frac{\eta}{\lambda d}. \tag{2.14}$$

The Fresnel approximation becomes

$$\begin{aligned} U(\nu, \mu) = & -\frac{\mathrm{i}}{\lambda d} \exp(\mathrm{i}kd) \exp[\mathrm{i}\pi\lambda d(\nu^2 + \mu^2)] \\ & \times \int_{-\infty}^{+\infty} \int_{-\infty}^{+\infty} I_{\mathrm{H}}(x,y) U_{\mathrm{R}}(x,y) \exp\left[\frac{ik}{2d}(x^2 + y^2)\right] \\ & \times \exp[-\mathrm{i}2\pi(x\nu + y\mu)] \,\mathrm{d}x \,\mathrm{d}y, \end{aligned} \tag{2.15}$$

which shows that the Fresnel approximation is the Fourier transformation of the term $I_{\mathrm{H}}(x,y)U_{\mathrm{R}}(x,y)\exp\left[\frac{ik}{2d}(x^2+y^2)\right]$.

Suppose that the hologram function $I_{\mathrm{H}}(x,y)$ is sampled on a rectangular grid of $N \times N$ points, Δx and Δy are the distances between neighboring pixels in horizontal and vertical direction. In accordance with the theory of Fourier transform, re-substitute

$$\Delta\xi = \frac{\lambda d}{N\Delta x}; \quad \Delta\eta = \frac{\lambda d}{N\Delta y} \tag{2.16}$$

in Eq. (2.15), and the discrete Fresnel transform is calculated as follows:

$$\begin{aligned} U(k,l) = & -\frac{\mathrm{i}}{\lambda d} \exp(\mathrm{i}kd) \exp\left[\mathrm{i}\pi\lambda d\left(\frac{k^2}{N^2\Delta x^2} + \frac{l^2}{N^2\Delta y^2}\right)\right] \\ & \times \sum_{m=0}^{N-1}\sum_{n=0}^{N-1} I_{\mathrm{H}} U_{\mathrm{R}}(m,n) \exp\left[\frac{ik}{2d}(m^2\Delta x^2 + n^2\Delta y^2)\right] \\ & \times \exp\left[-\mathrm{i}2\pi\left(\frac{km}{N} + \frac{ln}{N}\right)\right], \end{aligned} \tag{2.17}$$

where $k = 0, 1, \ldots, N-1$ and $l = 0, 1, \ldots, N-1$ are the discrete dimensions indices of the reconstructed image plane, and $n = 0, 1, \ldots, N-1$ and $m = 0, 1, \ldots, N-1$ are the discrete dimensions indices of the hologram plane. Similarly, one can obtain the discrete

Fraunhofer approximation:

$$U(k,l) = -\frac{\mathrm{i}}{\lambda d}\exp(\mathrm{i}kd)\exp\left[\mathrm{i}\pi\lambda d\left(\frac{k^2}{N^2\Delta x^2}+\frac{l^2}{N^2\Delta y^2}\right)\right] \times \sum_{m=0}^{N-1}\sum_{n=0}^{N-1} I_{\mathrm{H}}(m,n)U_{\mathrm{R}}(m,n)\exp\left[-\mathrm{i}2\pi\left(\frac{km}{N}+\frac{ln}{N}\right)\right]. \tag{2.18}$$

However, it is necessary to point out that the discrete Fourier transform (DFT) is restricted to a finite size whereas the continuous propagation is defined over an infinite plane. This problem may lead to aliasing artifacts, which can be limited by applying zero-padding to the planes before propagation. More information about zero-padding can be found in Liu (2012).

2.2.1.2 Angular Spectrum Method

Formulation of the angular spectrum method However, the above approximation methods have the deficiency of imprecision or inefficiency. Another method based on angular spectrum propagation is conducted without any approximation. The diffraction integral of Eq. (2.6) can be rewritten by a convolution as follows:

$$\begin{aligned} U(\xi,\eta) &= g(x,y) * h(\xi,\eta,x,y), \\ g(x,y) &= I_{\mathrm{H}}(x,y)U_{\mathrm{R}}(x,y), \end{aligned} \tag{2.19}$$

where the symbol $*$ represents the convolution operation, $h(\xi,\eta,x,y)$ is the propagation kernel which is given by

$$h(\xi,\eta,x,y) = -\frac{\mathrm{i}}{\lambda}\frac{\exp(\mathrm{i}kr)}{r}\cos\theta, \quad \cos\theta = \frac{d}{r}. \tag{2.20}$$

According to the convolution theorem, the Fourier transform of a convolution operation is the product of two individual Fourier transforms. By which,

$$\mathcal{F}\{U(\xi,\eta)\} = \mathcal{F}\{g(x,y)\}\mathcal{F}\{h(\xi,\eta,x,y)\} \tag{2.21}$$

is obtained, where $\mathcal{F}$ represents the Fourier transform. In Fourier spectrum, Eq. (2.19) is rewritten as

$$\mathcal{U}(\nu,\mu;d) = G(\nu,\mu;0)H(\nu,\mu;d), \tag{2.22}$$

where ν and μ are the Fourier frequencies in the x and y directions, respectively, and $\mathcal{U}$, G, and H represent the Fourier transforms of the corresponding functions. The transfer function $H(\nu, \mu; d)$ is given by

$$H(\nu, \mu; d) = \exp\left(i2\pi d\sqrt{\frac{1}{\lambda^2} - \mu^2 - \nu^2}\right). \tag{2.23}$$

Consequently, the reconstructed object field is formulated by the angular spectrum method as follows:

$$U(\xi, \eta) = \mathcal{F}^{-1}\{G(\nu, \mu; 0)H(\nu, \mu; d)\}, \tag{2.24}$$

where $\mathcal{F}^{-1}$ is the inverse Fourier transform.

Discrete reconstruction Just as with the setup for the approximation methods, both the hologram plane and real image plane are squares and sampled on a $N \times N$ grid. So the discretized object field at the real image plane can be evaluated as

$$U(k, l) = \frac{1}{N^2}\sum_{m=0}^{N-1}\sum_{n=0}^{N-1} G(m, n) \exp\left(i2\pi d\sqrt{\frac{1}{\lambda^2} - \left(\frac{m}{N}\right)^2 - \left(\frac{n}{N}\right)^2}\right) \times \exp\left[i2\pi\left(\frac{km}{N} + \frac{ln}{N}\right)\right], \tag{2.25}$$

where $G(m, n)$ represents the discrete Fourier transform of $g(x, y)$.

2.2.2 Computer-Generated Holograms

Computer-generated hologram (CGH) is a technique allowing one to create digital holograms by mathematical means of calculating the optical wave propagation. It is released from physical recording limitations by approximating the light or wave propagation with less complexity, as light propagation is affected by several phenomena like reflection, refraction, interference, diffraction, and so on.

Generally speaking, the basic stages for synthesizing a CGH can be listed as follows:

1. selecting a 3D object, real or virtual, and designing a mathematical model of the object;
2. defining the geometry of wave propagation from the object to the hologram plane;
3. computing the object wave at the hologram plane by carrying out the discrete transform on the representation of wave propagation; and
4. computing intensities of the interference patterns.

In CGH, both real and virtual objects can be chosen for recording. For real objects, it is possible to calculate holograms using their multiview projection images captured by a camera array. However, a significant advantage of CGH is that it is able to create digital holograms from virtual objects. To this end, CGH is appealing in 3D technology.

2.2.2.1 The State of the Art of CGH

For the past decades, techniques to create holograms using CGH have been developed, going from ray-tracing methods (Stein et al., 1992) with point source objects to the mainstream wave-oriented methods (Yamaguchi, 2011) with geometrical objects.

In the preliminary stage, a 3D object is considered to be composed of points which emit spherical waves as point sources. The object waves on the hologram plane are then calculated by tracing the ray from a point source on the object to a sampling point on the hologram. This method is the most flexible for synthesizing holograms, thanks to its simple principle (explained in the following). However, not all object points are at the same distance from the hologram plane, therefore the computation cost could be rather enormous. Many methods have been attempted to reduce the computation cost, such as by using look-up tables (Lucente, 1993), PC hardware (Ritter et al., 1999; Bove et al., 2005), special central processing units (CPU) (Ito et al., 1996; Shimobaba et al., 2000, 2002; Ito et al., 2005), and graphics processing units (GPU) (Masuda et al., 2006). It thus becomes possible to handle a huge number of point source objects, thanks to the progress of high-speed calculation techniques.

Conversely, wave-oriented methods follow the diffraction theory to calculate the object fields scattered from each object surface, propagating to the hologram plane. The geometrical objects are represented as compositions of elementary diffracting elements, for example, interpolated projection images or multiview images of real or virtual objects, virtual objects constituted of polygonal meshes, or point cloud objects.

In Abookasis and Rosen (2006), holograms are synthesized from a set of projection images of a 3D object rendered from multiple perspectives (angulars). Differently, an initial hologram is defined first to be used as a holographic transparency. It is a complex matrix containing the wavefront distribution on the Fourier plane. This matrix can be converted to different types of holograms by computing the corresponding wave propagation. Similar works are also conducted in Sando et al. (2003) and Shaked et al.

(2009), where an array of cameras is usually used. In order to make the system more compact or even portable, the utilization of a single camera with external lens or a plenoptic camera with a micro lens array inside has been reported in Park et al. (2009), Mishina et al. (2006), Lee et al. (2013), and Shaked and Rosen (2008). All these systems synthesize holograms in similar ways, with the main difference in the capture of light ray or wavefront distribution.

A wave-oriented polygon-based method was first proposed in Matsushima (2005). The main advantage of using a polygonal represented object is that a fast Fourier transform (mainly by angular spectrum methods) can be applied to wavefront propagation for each planar surface. Each polygon is considered as a surface source of light. The object field at the hologram plane is computed as the sum of the necessary transformations of each surface field. Some details will be given in the following. Some improvements based on the basic idea in Matsushima (2005) have been proposed in order to reduce computational complexity, including hidden surface removal or occluded scene reconstruction (Matsushima and Kondoh, 2004; Ahrenberg et al., 2008; Kim et al., 2008; Matsushima and Nakahara, 2009).

Recently, methods combining ray-tracing and wave-oriented methods have become popular. In Chen and Wilkinson (2009), the point emitter is selected as the geometric primitive. Full-parallax holograms are generated based on the Rayleigh-Sommerfeld diffraction theory by using angular spectra. In Wakunami and Yamaguchi (2011), a new algorithm for calculating CGH using ray-sampling plane is introduced, enabling one to reproduce a high-resolution image for deep 3D scene with angular reflection properties. The ray-sampling plane is defined parallel to the hologram plane and set close to the object. The light rays emitted by the object are first sampled at the ray-sampling plane, and then transformed into the wavefront by using Fourier transform. At the end, by applying a Fresnel diffraction for wavefront propagation, the wavefront on the hologram plane is obtained.

2.2.2.2 Methods for CGH

A simple introduction to ray-tracing and wave-oriented methods is given in the following based on in-line holography.

Ray-tracing methods Ray-tracing algorithms are the most flexible methods for synthesizing holograms thanks to their simplicity. Based on the isotropic point source model, the complex amplitude at one point of

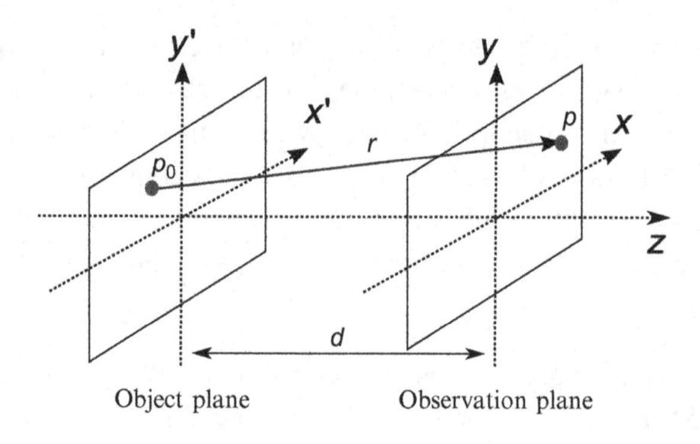

Figure 2.4 Coordinate system of wave propagation.

the hologram plane is the sum of complex amplitudes of all radiations coming from each object point. Given the coordinate system of wave propagation by ray-tracing methods in Fig. 2.4, each point P_0 on the object surface can be considered as a source of a spherical wave which can be represented as

$$E(P_0) = A_0 \exp(i\varphi_0), \tag{2.26}$$

where $E(P_0)$ is the complex amplitude with the real-valued amplitude A_0 and the initial phase φ_0 at point P_0. With the design of $\varphi_0 = 0$, the wave representation at point P in space can be approximated by satisfying the scalar Helmholtz equations as

$$E(P) = A_0 \frac{\exp(ikr)}{r}, \tag{2.27}$$

where $E(P)$ is the complex amplitude of the wave at point P generated by point P_0, and r is the distance between point P and point P_0. Then, the object wave at the hologram plane is represented by

$$U(x,y) = \int_{-\infty}^{+\infty} \int_{-\infty}^{+\infty} A(x',y') \frac{\exp(ikr)}{r} \, dx' \, dy'. \tag{2.28}$$

In the discrete form, we obtain

$$U(m,n) = \sum_{k=0}^{N-1} \sum_{l=0}^{N-1} A(k,l) \frac{\exp(ikr_{mk,nl})}{r}, \tag{2.29}$$

where $k = 0, 1, \ldots, N-1$ and $l = 0, 1, \ldots, N-1$ are the discrete dimensions indices of the object plane, $n = 0, 1, \ldots, N-1$ and $m = 0, 1, \ldots, N-1$ are the discrete dimensions indices of the hologram plane, and $r_{mk,nl}$ is the distance between the pixel (m,n) in the hologram plane and the pixel (k, l) in the object plane. Consequently, the intensity of the recorded holograms can be expressed by

$$I_{\mathrm{H}}(m,n;\varphi) = |U_{\mathrm{R}}(\varphi) + U(m,n)|^2. \tag{2.30}$$

Wave-oriented methods

Since wave-oriented methods are quite popular for polygon-based objects, the description hereafter is based on them.

Wave-oriented methods describe the light propagation by following diffraction theory, either the Fresnel-Kirchoff formulation or the Rayleigh-Sommerfeld one. Different from the isotropic radiation setup in ray-tracing methods, wave-oriented methods treat the radiation from the elemental surface in the forward direction with less side radiation and no back radiation. It consequently results in less diffusive object surface than the ray-tracing approach. To this end, the first mission for generating holograms from complex geometrical objects is to define a suitable surface properties function, including brightness, shape, and diffusion.

Different from ray-tracing methods, each object surface in wave-oriented methods has its own local coordinates $(\hat{x}, \hat{y})$, as illustrated in (Fig. 2.5). The properties function of the nth surface can be generally defined as

$$U_n(\hat{x}_n, \hat{y}_n) = a_n(\hat{x}_n, \hat{y}_n)\Psi_n(\hat{x}_n, \hat{y}_n), \tag{2.31}$$

where $a_n(\hat{x}_n, \hat{y}_n)$ is the real amplitude representing the shape and texture information of the surface, and $\Psi_n(\hat{x}_n, \hat{y}_n)$ is a numerical

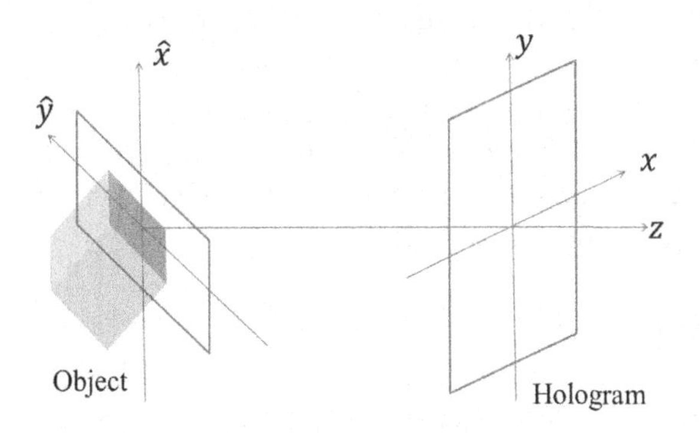

Figure 2.5 Geometry of wave-oriented methods with tilted object surface.

diffuser given by

$$\Psi_n(\hat{x}_n, \hat{y}_n) = \exp[i\psi_n(\hat{x}_n, \hat{y}_n)], \tag{2.32}$$

where $\psi_n(\hat{x}_n, \hat{y}_n)$ is a phase. However, neither constant phase for all surfaces nor full random phases are appropriate to the diffusive phase. The former one lacks diffusiveness, while the latter one causes a large Fourier frequency. A numerical diffuser used for $\psi_n(\hat{x}_n, \hat{y}_n)$ has been reported in Bräuer et al. (1991) for Fourier holograms.

The propagation direction of the light, given only by the product of the amplitude $a_n(\hat{x}_n, \hat{y}_n)$ and phase $\Psi_n(\hat{x}_n, \hat{y}_n)$, is perpendicular to the nth surface. It means the polygon field may not reach the hologram plane sufficiently if the local coordinate is not parallel to the hologram plane. As a result, the center of the surface function spectrum should be shifted so that the light propagates nearly perpendicularly to the hologram along Z-axis. This step is defined as Fourier spectrum remapping. The shift operation is related to a rotation transformation $\mathcal{R}$ (Matsushima, 2005), which can be written as

$$U_n(x_n, y_n) = \mathcal{R}\{U_n(\hat{x}_n, \hat{y}_n)\}. \tag{2.33}$$

The complex field superimposed at the hologram plane from all surfaces is consequently able to be obtained by applying the diffraction theories (introduced in last section) to each surface field $U_n(x_n, y_n)$.

2.3 PHASE-SHIFTING DIGITAL HOLOGRAPHY

Phase-shifting interferometry (PSI) was proposed to improve reconstruction quality in DH. The complex amplitude in the hologram plane can be completely calculated, thus the complex amplitude in any plane, for example, the virtual or real image plane, can be reconstructed by using the Fresnel-Kirchhoff diffraction formulation. The object image can consequently be reconstructed without spurious images. In this section, the motivation of applying phase-shifting digital holography (PSDH) and the principle of the phase-shifting algorithm are described.

2.3.1 Motivation

For holograms recorded by the geometry shown in Fig. 2.1, a bright square would appear in the center of the reconstructed image and cover the object image as shown in Fig. 2.6. It is called zero-order image due to the

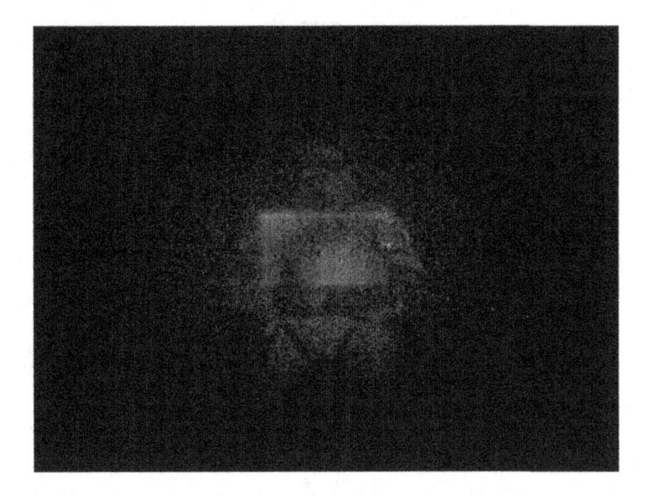

Figure 2.6 Example of reconstructed object with zero-order image.

undiffracted reconstruction wave which is expressed by the first two terms in Eq. (2.2). This type of setup is called in-line setup, which means there is no offset angle between the object wave and the reference wave when they are introduced to the hologram plane. On the other hand, with the same geometry, the real image and virtual image are located in one line as shown in Fig. 2.2. Normally only one image is visible in the reconstruction; however, in some cases, for example, if the recording distance is very short, the other out of focus image (twin image) might disturb the reconstructed one. For these two reasons, Leith and Upatnieks proposed an off-axis setup where a tilted reference wave is introduced (Leith and Upatnieks, 1962), by which the reconstructed image is spatially separated from the zero-order image and twin image. However, this setup requires much higher spatial frequencies on the charged coupled device (CCD) than in-line setup. Meanwhile, in order to alleviate this problem, Yamaguchi and Zhang (1997) proposed PSDH based on in-line setup.

2.3.2 Principle of Phase-Shifting Digital Holography

The basic geometry of PSDH is shown in Fig. 2.7. Similar to the classic setup shown in Fig. 2.1, one branch of the split laser beam illuminates the object and interferes with the other beam (the reference beam) at the CCD device. The reference beam is reflected at the piezoelectric transducer (PZT) mirror that phase modulates the beam. By shifting a constant phase to the reference beam, different holograms are obtained to derive the complex amplitude of the object wave.

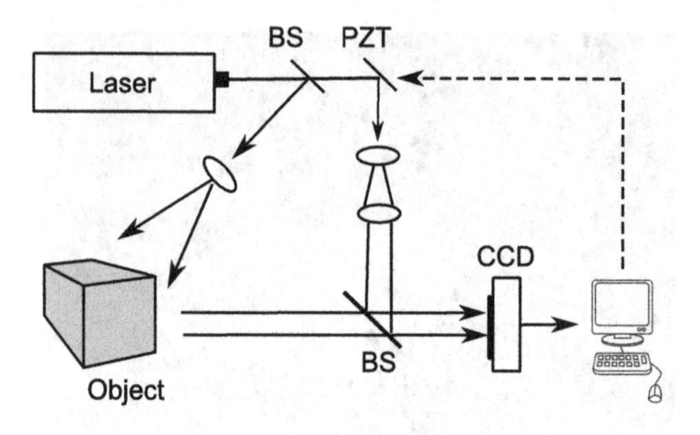

Figure 2.7 Setup for phase-shifting digital holography: PZT, piezoelectric transducer mirror.

In the simplest application of PSDH, the initial phase φ_R of the reference wave is zero and changes by $\pi/2$ at each step, the intensity of the interference patterns expressed in Eq. (2.2) is then updated as

$$I_H(x,y;\varphi_R) = |U_R(x,y;\phi) + U_O(x,y)|^2, \quad \varphi_R \in \left\{0, \frac{\pi}{2}, \pi\right\}. \tag{2.34}$$

The complex amplitude of the object wavefront at the hologram plane can be calculated directly using recorded holograms in the case of a three-step algorithm:

$$\hat{U}_O(x,y) = \frac{1-\imath}{4U_R^*}\left\{I_H(x,y;0) - I_H\left(x,y;\frac{\pi}{2}\right) + \imath\left[I_H\left(x,y;\frac{\pi}{2}\right) - I_H(x,y;\pi)\right]\right\}, \tag{2.35}$$

where U_R^* is the conjugate of the reference wave with $\varphi_R = 0$. Then the complex amplitude in the image plane, $I_H(x,y)$, can be reconstructed by using the Fresnel-Kirchhoff diffraction formulation.

CHAPTER 3

Basic Compression Tools

In this chapter, we first briefly introduce basic compression tools. Generally, compression techniques are classified into two categories: lossy and lossless techniques. A typical image compression scheme incorporates three fundamentals steps: transformation, quantization, and entropy coding, as shown in Fig. 3.1. Indeed, the JPEG 2000 image coding framework is commonly used for holographic compression because of its modular and extendable architecture. These three steps will be discussed in more details hereafter. We shall especially focus on wavelet-based schemes, as they offer appealing features for holographic data. Furthermore, as a special case of wavelet-based schemes, we shall also introduce JPEG 2000, which has been efficiently applied for the compression of holograms.

3.1 TRANSFORMATION: WAVELET-BASED TRANSFORM

In compression, the transformation step aims at a decorrelation of the data, compacting in this way the energy of the signal into few significant transform coefficients. The wavelet transform has been widely adopted in the field of image compression for its good property of spatial-frequency localization, which is also essential for compressing holographic data. The classical 1D discrete wavelet transform (1D-DWT) can be simply summarized as applying a pair of low-pass and high-pass filters to the input signal, followed by a downsampling of each filter output. A multiresolution decomposition can be obtained by further decomposing the low-frequency subband. For an image, a separable DWT is commonly used by applying the 1D-DWT once along the rows and once along the columns. Consequently, four subbands can be obtained by one decomposition: an approximation subband named LL and three detail subbands named HL, LH, and HH

Digital Holographic Data Representation and Compression. http://dx.doi.org/10.1016/B978-0-12-802854-4.00003-3

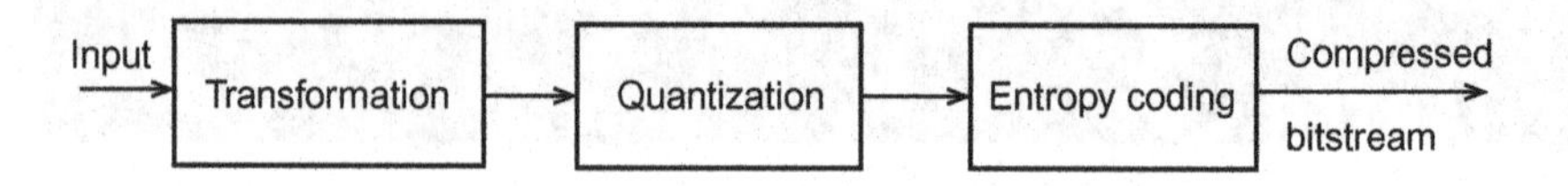

Figure 3.1 Generic compression scheme.

corresponding, respectively, to the horizontal, vertical, and diagonal orientations. Moreover, many different filters can be used—for example, the simplest 5/3 filter bank, has the "perfect reconstruction" property while the well-known Daubechies 9/7 filter bank has high compression efficiency (Daubechies, 1992).

3.2 QUANTIZATION

The quantization step is a lossy process which reduces the precision of the data samples into a discrete set of levels. The general quantization process is shown in Fig. 3.2.

Generally speaking, two kinds of quantization methods can be distinguished, based on uniform and adaptive quantizers. Moreover, quantization can operate on scalar or vector data samples, referred to as scalar quantization (SQ) and vector quantization (VQ), respectively.

Uniform scalar quantization The simplest example of SQ is uniform scalar quantization (USQ), by which an input sample x_i is mapped to a quantized value $\hat{x}_i$ by

$$\hat{x}_i = \left\lfloor \frac{x_i}{Q_{\text{step}}} \right\rfloor Q_{\text{step}}, \tag{3.1}$$

where Q_{step} is the quantization step size and $\lfloor \cdot \rfloor$ is the rounding operator.

Adaptive scalar quantization Adaptive scalar quantization (ASQ) has been proposed to reduce quantization errors for input values which are

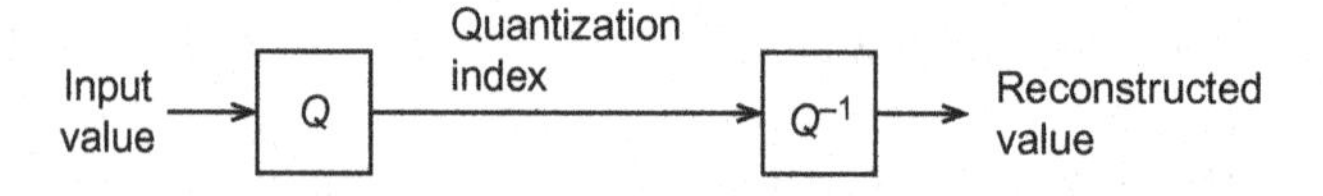

Figure 3.2 Quantization process: Q represents the encoding or quantization stage and Q^{-1} represents the decoding or inverse quantization stage.

not uniformly distributed. In this context, the Lloyd-Max algorithm (Max, 1960; Lloyd, 1982) aims at minimizing the mean squared error (MSE) resulting from quantization.

Vector quantization In the case of VQ, the Linde-Buzo-Gray algorithm (LBG-VQ) (Linde et al., 1980) is also designed to iteratively minimize MSE, but on multidimensional inputs. Two criteria should be satisfied: (1) the nearest neighbor condition, in other words, all vectors that are closer to a specific centroid than any other centroids, should be contained in one region and approximated to this specific centroid as output; and (2) a centroid should be the average of all the vectors in one region.

3.3 ENTROPY CODING

After the transformation and quantization steps, the quantized transform coefficients need to be encoded, in other words, converted to a binary codestream. Basically, the principle of entropy coding is to assign variable length codewords (VLC) to input symbols to exploit the nonuniform statistical distribution of these symbols, hence resulting in an overall reduced number of bits.

As previously stated, wavelet-based schemes are especially interesting. In this case, the property of quality scalability can effectively be achieved. For this purpose, the coding scheme should encode the coefficients in decreasing order of their importance with reduced accuracies. During the decoding procedure, the decoded image is first approximated by the most significant coefficients, then gradually refined by the less significant ones. The development of wavelet-based embedded codecs has been the subject of numerous research works and readers can find an overview of the corresponding state of the art in Bildkompression and der Technik (2003).

The embedded zerotree wavelet (EZW) coding (Shapiro, 1993) and its extended version named set partitioning in hierarchical trees (SPIHT) (Said and Pearlman, 1996) have been proven to be very effective for image compression. Particularly, SPIHT has been applied for compressing holographic data given its good performance. Both schemes are based on three principles: partial ordering by magnitude with a set partitioning sorting algorithm, ordered bit plan transmission, and exploitation of cross-scale similarities of the wavelet coefficients. However, they differ in the way

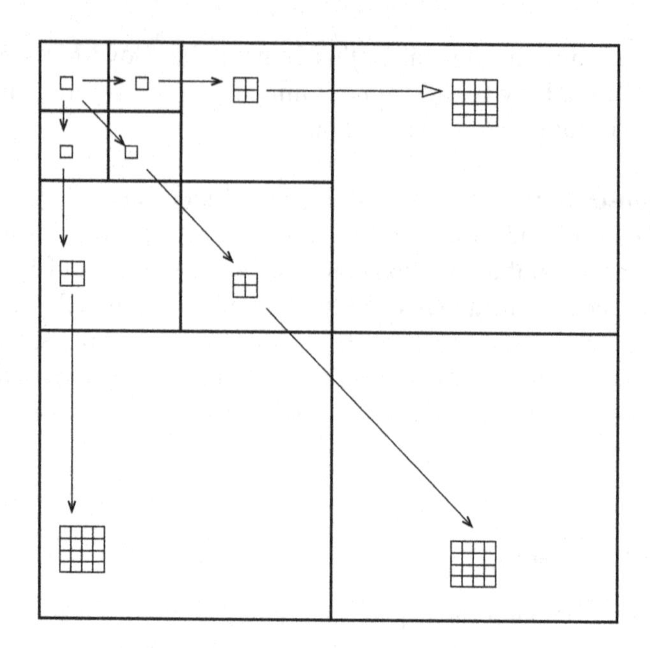

Figure 3.3 Zerotree or spatial-orientation tree.

subsets of coefficients are partitioned. Basically, the zerotree concept based on raster scan in EZW is improved by using sorted lists in SPIHT instead.

The basic concepts used in SPIHT are now briefly introduced. Figure 3.3 shows the definition of the zerotree or spatial-orientation tree. The arrows are oriented from the parent node to the offspring nodes. Each node of the tree corresponds to a pixel and has either zero or four offsprings with same spatial orientation in the next finer level. In practical implementation, four kinds of subsets are used: one set containing all the roots and three other sets, respectively, containing all the offsprings, descendants, and indirect descendants of the entry node. Three ordered lists are used to decide the order in which the subsets are tested for significance, which are called list of insignificant sets (LIS), list of insignificant pixels (LIP), and list of significant pixels (LSP), respectively. The descendant sets of the roots are added in LIS for initialization.

SPIHT aims at improving the zerotree concept by replacing the raster scan with sorted lists. During the sorting pass, the pixels in LIP are tested and moved to LSP if they are significant. Sets in LIS are evaluated following the LIS order. A set is removed from LIS when it is found to be significant. New subsets with more than one element are added back to the LIS and the element of the removed sets are added to the LIP or LSP depending on their significance. The pixels in LSP are refined with an additional bit in the refinement pass. The two passes are iterated until the threshold for testing the significance becomes lower than the minimum wavelet coefficient.

3.4 JPEG 2000

JPEG 2000 is the state-of-the-art standard for still image coding (Taubman and Marcellin, 2001; Schelkens et al., 2009). On top of very high coding efficiency, JPEG 2000 also provides with a number of highly desirable features such as seamless progressive transmission by resolution or quality, lossy to lossless compression, random code stream access, continuous-tone and bi-level compression, and region of interest. It has been widely adopted in several application domains, including digital cinema and broadcasting environments, medical imaging, surveillance, and cultural heritage archives.

Figure 3.4 shows the typical schematic of JPEG 2000, which consists of five main steps: image tiling, wavelet transform, quantization, embedded code-blocks encoding, and rate-distortion optimization. Image tiling is useful to handle very large images. DWT is then applied. Two wavelet transforms are defined in the core coding system: the irreversible floating-point CDF 9/7 transform kernel and the reversible integer 5/3 transform kernel. Each subband is then subdivided into code-blocks, which are independently encoded. In turn, each code-block is quantized and entropy coded in a bitplane-by-bitplane process using a context-based adaptive binary arithmetic encoding. Before assembling the individual bitstreams into a single, final bitstream, each code-block bitstream is truncated by the Lagrangian rate-distortion optimal technique. The truncated bitstreams are then concatenated together to form the final bitstream. This technique is named embedded block coding with optimal truncation (EBCOT) (Taubman, 2000).

The code-block structure facilitates random access in the data, a property which can often be useful. Moreover, unlike most alternative coding

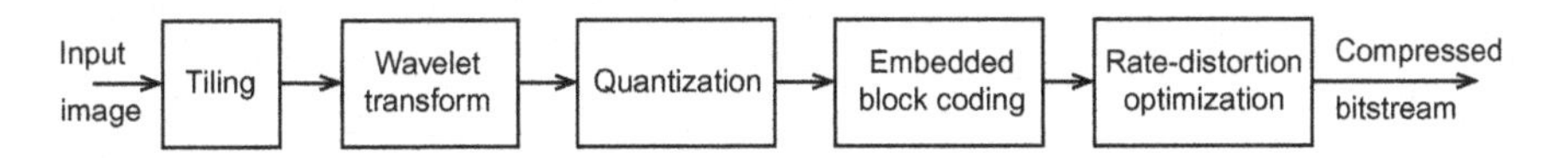

Figure 3.4 Schematic block diagram of JPEG 2000.

schemes, JPEG 2000 compression can be both lossy and lossless, always produces an intrinsically progressive bitstream, and can compress a region of interest with higher quality. Finally, the versatility of the JPEG 2000 encoding architecture makes it an appealing solution when considering the compression of holographic data.

CHAPTER 4

Holographic Data Representations

In order to design an effective compression method for a specific type of data, a good understanding of its statistical characteristics is necessary. For digital holographic data, different representations are possible, as we shall see in this chapter. These representations are equivalent in the sense that they all reconstruct the same object. However, different representations may have different statistical characteristics. Therefore, we also investigate their suitability for compression.

4.1 REPRESENTATIONS OF DIGITAL HOLOGRAPHIC DATA

Hereafter, we consider more specifically the case of phase-shifting digital holography (PSDH), as the phase-shifting algorithm can effectively be used for the recording steps in order to obtain reconstructed object images with good quality. However, having three sets of intensity data is not a good target for compression. Therefore, reduced datasets are desirable to mitigate the compression burden. On the other hand, the data should keep all the useful information for reconstruction.

According to the phase-shifting algorithm, the most important information for reconstruction is the complex object field at the hologram plane. If this field can be obtained or expressed by a reduced amount of information, this information can be considered as a representation of interference patterns. So far, a few representations have been mainly used in accordance with the phase-shifting interferometry algorithm, which are introduced in the following.

Digital Holographic Data Representation and Compression. http://dx.doi.org/10.1016/B978-0-12-802854-4.00004-5

4.1.1 Intensity-Based Representation

From Eqs. (2.34) and (2.35), the intensity-based representations can be equivalently expressed in two ways.

Intensity information The intensity information, $I_{\mathrm{H}}(x, y; \varphi_R), \varphi_{\mathrm{R}} \in \left\{0, \frac{\pi}{2}, \pi\right\}$, is the direct representation. However, compressing three sets of data are not optimal in terms of compression.

Shifted distance information To this end, and observing that only two differences of intensity terms are needed in Eq. (2.35), we introduce the difference data $D^{(1)}$ and $D^{(2)}$ given by:

$$\begin{cases} D^{(1)}(x,y) = I_{\mathrm{H}}(x,y;0) - I_{\mathrm{H}}\left(x,y;\frac{\pi}{2}\right) \\ D^{(2)}(x,y) = I_{\mathrm{H}}\left(x,y;\frac{\pi}{2}\right) - I_{\mathrm{H}}(x,y;\pi)\,. \end{cases} \tag{4.1}$$

In this case, only two sets of data are necessary to reconstruct the complex field in Eq. (2.35). Since the raw data rate is reduced and it can directly be obtained from the difference of intensity information, the definition of $D^{(1)}$ and $D^{(2)}$ is highly optimal as one representation of interference patterns. As this representation is defined by the difference of phase-shifted holograms, hence, we referred to it as "shifted distance information."

4.1.2 Complex Amplitude-Based Representation

Complex amplitude-based representations are derived from different expressions of the complex field. Noted that a complex number has two kinds of expressions, one is defined in Cartesian coordinate system and the other one in polar coordinate system.

Real-imaginary information In Cartesian coordinate system, a complex number can be expressed in the form $a + bi$, where a and b are the real part and imaginary part, respectively. Thus, the complex field obtained by Eq. (2.35) can be expressed as:

$$\hat{U}_{\mathrm{O}}(x,y) = \Re(\hat{U}_{\mathrm{O}}(x,y)) + \mathrm{i} \cdot \Im(\hat{U}_{\mathrm{O}}(x,y)), \tag{4.2}$$

where $\Re(\hat{U}_{\mathrm{O}})$ and $\Im(\hat{U}_{\mathrm{O}})$ are, respectively, the real and imaginary parts of the complex object field at the hologram plane.

Amplitude-phase information An alternative way to express a complex number is by polar coordinate system, where the complex field can be written by Euler's formula as

$$\hat{U}_O(x,y) = \hat{A}_O(x,y) \cdot \exp[i\hat{\psi}_O(x,y)], \tag{4.3}$$

where

$$\hat{A}_O(x,y) = |\hat{U}_O(x,y)| = \sqrt{\Re^2(\hat{U}_O(x,y)) + \Im^2(\hat{U}_O(x,y))}, \tag{4.4}$$

$$\hat{\psi}_O(x,y) = \begin{cases} \arctan \dfrac{\Im(\hat{U}_O(x,y))}{\Re(\hat{U}_O(x,y))} & \text{if } \Re(\hat{U}_O(x,y)) > 0 \\ \arctan \dfrac{\Im(\hat{U}_O(x,y))}{\Re(\hat{U}_O(x,y))} + \pi & \text{if } \Re(\hat{U}_O(x,y)) < 0 \text{ and } \Im(\hat{U}_O(x,y)) \geq 0 \\ \arctan \dfrac{\Im(\hat{U}_O(x,y))}{\Re(\hat{U}_O(x,y))} - \pi & \text{if } \Re(\hat{U}_O(x,y)) < 0 \text{ and } \Im(\hat{U}_O(x,y)) < 0 \\ \dfrac{\pi}{2} & \text{if } \Re(\hat{U}_O(x,y)) = 0 \text{ and } \Im(\hat{U}_O(x,y)) > 0 \\ -\dfrac{\pi}{2} & \text{if } \Re(\hat{U}_O(x,y)) = 0 \text{ and } \Im(\hat{U}_O(x,y)) < 0. \end{cases} \tag{4.5}$$

The 3D information of the recorded object is fully contained in this representation, especially the phase information, which has a significant influence on reconstruction.

The above representations are equivalent, up to the precision of computation, in the sense that they all reconstruct the same object. The intensity of the interference patterns, Eq. (2.34), can be rewritten as

$$I_H(x,y;\varphi_R) = A_R^2(x,y) + A_O^2(x,y) + 2A_R(x,y)A_O(x,y)\cos(\varphi_R - \varphi_O). \tag{4.6}$$

It can be noticed that the third term contains all effective information in the sense that the first and second terms will be eliminated during the subtraction operations in Eq. (2.35). Thus, by only retaining the third term, three signals can be defined as follows:

$$\begin{aligned} I^{(1)}(x,y) &= 2A_R(x,y)A_O(x,y)\cos(0 - \varphi_O(x,y)) \\ &= 2A_R(x,y)A_O(x,y)\cos(\varphi_O(x,y)) \\ &= 2A_R(x,y)\Re(U_O(x,y)) \\ I^{(2)}(x,y) &= 2A_R(x,y)A_O(x,y)\cos\left(\frac{\pi}{2} - \varphi_O(x,y)\right) \\ &= 2A_R(x,y)A_O(x,y)\sin(\varphi_O(x,y)) \\ &= 2A_R(x,y)\Im(U_O(x,y)) \\ I^{(3)}(x,y) &= 2A_R(x,y)A_O(x,y)\cos(\pi - \varphi_O(x,y)) \\ &= -I^{(1)}(x,y). \end{aligned} \tag{4.7}$$

Note that $D^{(1)}$ and $D^{(2)}$ given by Eq. (4.1) can be re-expressed as follows:

$$\begin{cases} D^{(1)}(x,y) = I^{(1)}(x,y) - I^{(2)}(x,y) \\ D^{(2)}(x,y) = I^{(1)}(x,y) + I^{(2)}(x,y). \end{cases} \tag{4.8}$$

Therefore, we can be assured that real-imaginary information has similar features as shifted distance information.

Selecting a suitable representation for digital holographic data is very important for further processing, for example, compression, transmission, and storage. Indeed, in order to design an effective compression method for a specific type of data, a good understanding of its statistical characteristics is necessary. In this context, it is important to address the following questions:

1. Which representation is more suitable for compression? What are its main statistical characteristics?
2. For a given representation, does one of the information components affect more the reconstruction quality?
3. How to design a compression scheme? Should it be specifically designed for a given representation?

A thorough study and analysis is needed in order to characterize different holographic data representations and to start addressing the above issues. This is the subject of the two following sections.

4.2 STUDY OF PROBABILITY DISTRIBUTIONS

In this section, we investigate the probability distributions for each holographic data representation.

Three virtual objects ("Luigi," "Girl," and "Bunny") shown in Fig. 4.1 are selected in the experiments, and the corresponding holograms are computer-generated using PSDH based on the methods described in Chapter 2.

Examples of the three representations: shifted distance information, real-imaginary information, and amplitude-phase information, obtained from the same virtual object, are shown in Figs. 4.2–4.4, respectively. First, we can observe that the patterns in the two components of the shifted distance information are visually similar. The same observation also holds for the

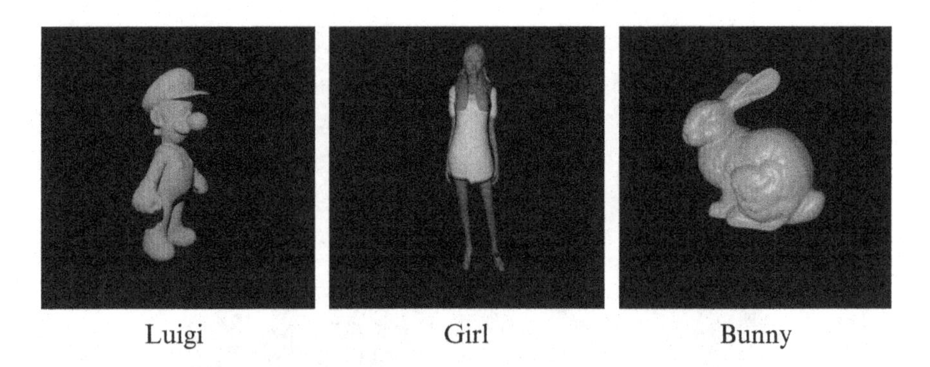

Figure 4.1 Test objects: "Luigi," "Girl," and "Bunny."

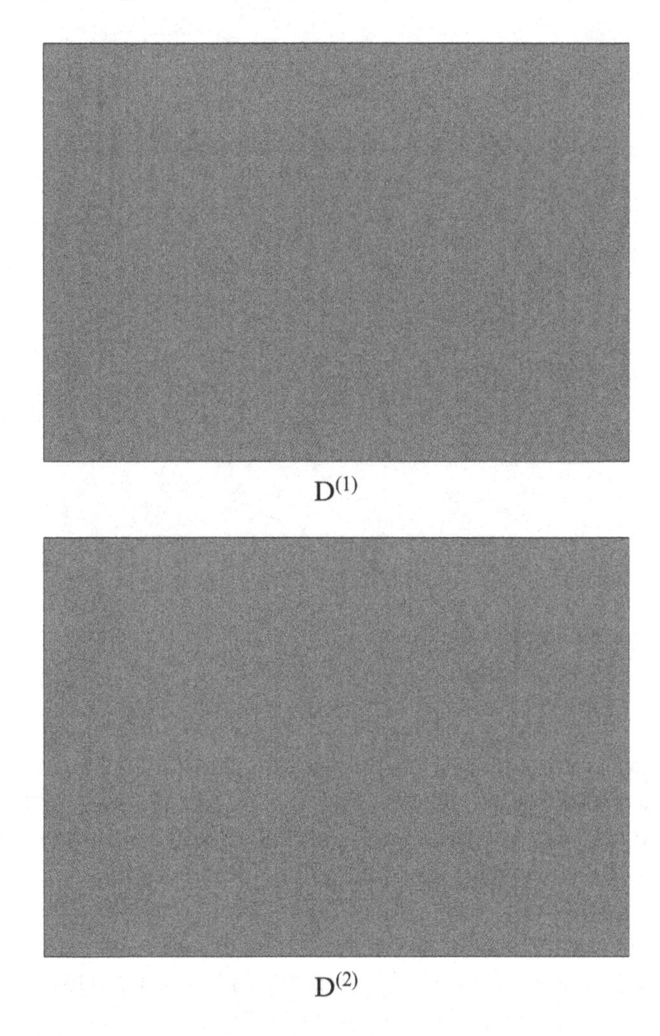

Figure 4.2 Example of shifted distance representation $D^{(1)}$ and $D^{(2)}$.

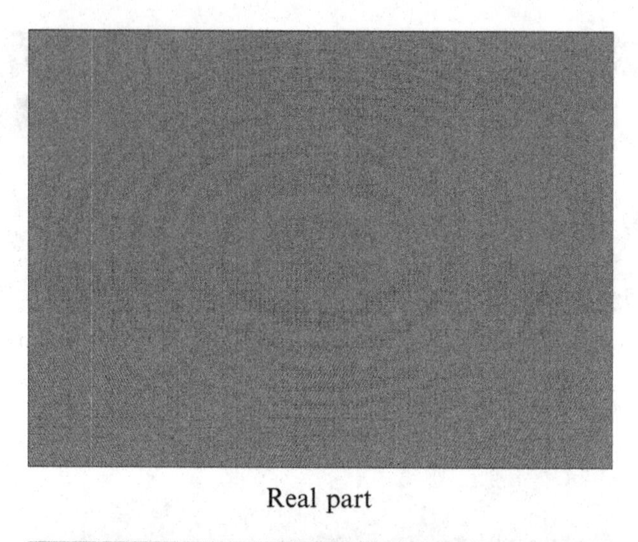

Real part

Imaginary part

Figure 4.3 Example of real-imaginary representation.

real-imaginary information. However, in the amplitude-phase representation, the two patterns are rather dissimilar. It can be inferred that more inter-component redundancies exist in shifted distance information or real-imaginary information than in the amplitude-phase information. As a consequence, compression methods exploiting inter-component redundancies are expected to be more effective in the shifted distance and read-imaginary cases, but rather ineffective in the amplitude-phase case. However, these observations need to be quantitatively verified by experiments.

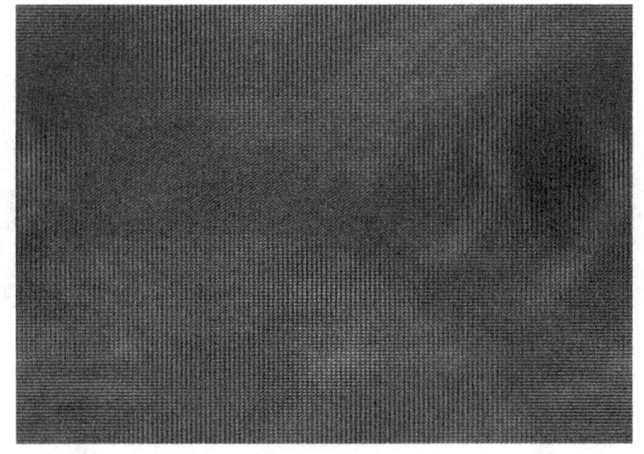

Phase

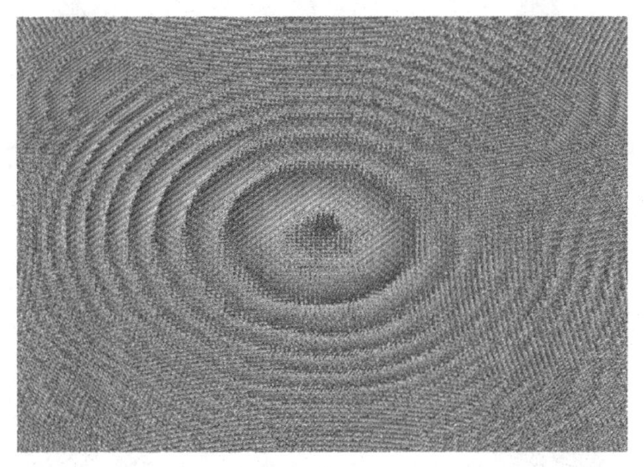

Amplitude

Figure 4.4 Example of amplitude-phase representation.

4.3 COMPARATIVE STUDY OF DIFFERENT REPRESENTATIONS

In this section, comparative studies of compressing different holographic data representations with various quantization methods are described, following the work in Xing et al. (2014b). More specifically, the impact of components on the reconstruction quality is first investigated. Then, redundancies across components are analyzed.

Next, probability distributions for each representation are studied, as shown in Fig. 4.5. It is straightforward that, except for the phase component,

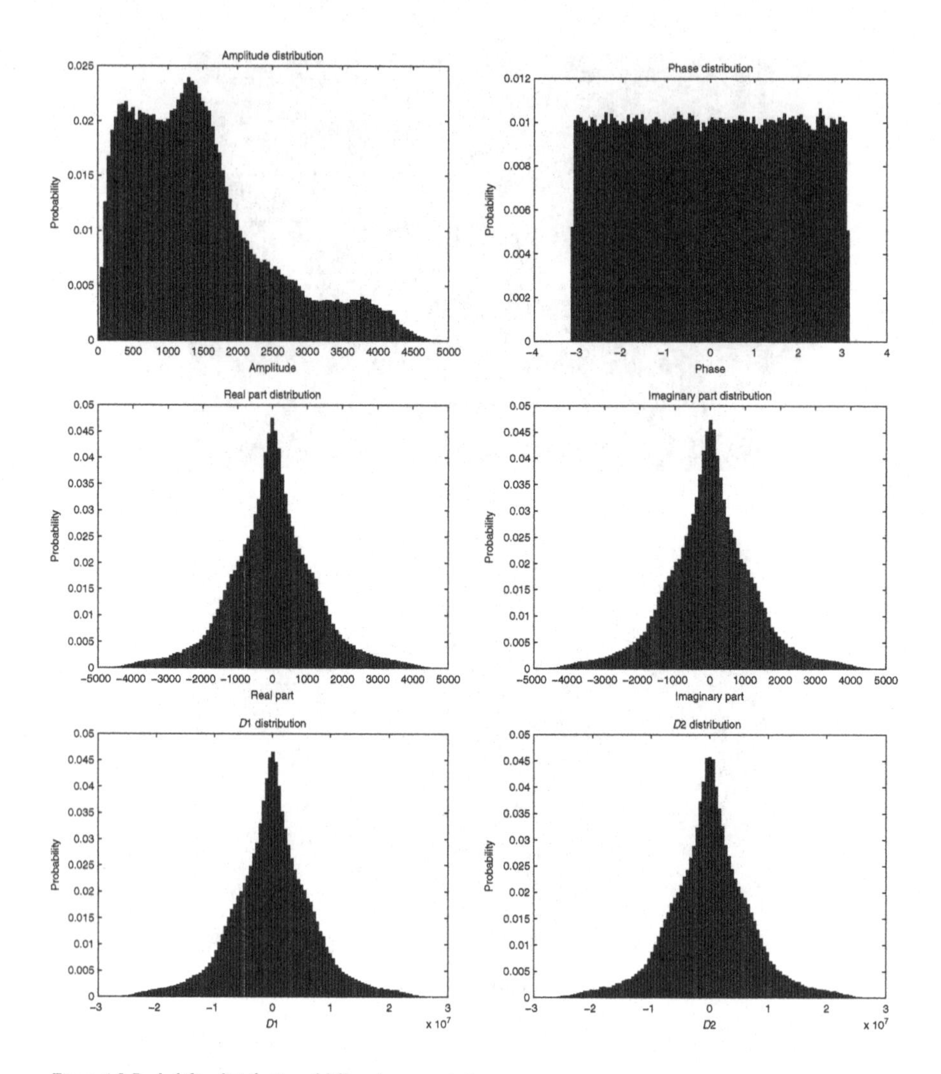

Figure 4.5 Probability distribution of different representations.

all other components have clear nonuniform distributions. Besides, the real-imaginary components and the shifted distance components share very similar distributions.

Hereafter, some common methods of quantization tools are used to study and analyze the characteristics of different holographic data representations. In particular, SQ processes input components independently, whereas VQ processes them jointly.

Impact analysis of each component on reconstruction quality We first investigate the impact of each component on the quality of the reconstructed object. For this purpose, each component is individually quantized using scalar quantization, and the corresponding object is then reconstructed.

For the sake of simplicity, only the peak signal-to-noise ratio (PSNR) index is reported hereafter, whose definition is given as follows:

$$\text{PSNR} = 10\log_{10}\left(\frac{\max(I_{\text{ref}})^2 - \min(I_{\text{ref}})^2}{\text{MSE}}\right),$$
$$\text{MSE} = \frac{1}{MN}\sum_{m=1}^{M}\sum_{n=1}^{N}[I_{\text{ref}}(m,n) - I_{\text{rec}}(m,n)]^2, \quad (4.9)$$

where I_{ref} is the $M \times N$ reference image (or the image reconstructed from uncompressed holographic data), and $\max(I_{\text{ref}})$ and $\min(I_{\text{ref}})$ are, respectively, the maximum and minimum pixel values of the reference image. I_{rec} is the approximated image reconstructed from compressed holographic data and MSE is the mean square error. The reference and approximated images correspond to the reconstructed intensity rather than amplitude. Comparative studies, experimental results, and discussions are described hereafter.

In the first set of experiments, three representations are separately quantized by uniform scalar quantization (USQ). More precisely, each component is quantized with bit levels from 2 to 7. Tables 4.1–4.3 evaluate the reconstruction quality of the object "Luigi." Figures 4.6–4.8 show corresponding illustrations.

Table 4.1 PSNR Obtained for "Luigi" by Quantizing Amplitude-Phase Information Using USQ (Am: Amplitude, Ph: Phase)

Number of Bits		Ph 2	3	4	5	6	7
Am	2	24.66	30.79	31.27	31.22	31.26	31.27
	3	25.14	35.06	38.45	39.04	39.47	39.58
	4	25.24	36.43	42.71	44.79	46.76	47.34
	5	25.25	36.67	44.00	47.29	51.89	54.15
	6	25.26	36.72	44.25	47.85	53.80	58.10
	7	25.26	36.73	44.30	47.97	54.38	59.80

Table 4.2 PSNR Obtained for "Luigi" by Quantizing Real-Imaginary Information Using USQ (Re: Real Part, Im: Imaginary Part)

Number of Bits		Im 2	3	4	5	6	7
Re	2	20.24	25.30	26.70	26.91	26.93	26.94
	3	25.26	32.05	35.50	36.27	36.42	36.47
	4	26.63	35.51	40.42	42.70	43.38	43.56
	5	26.82	36.29	42.67	46.74	48.77	49.50
	6	26.85	36.46	43.37	48.83	52.75	54.85
	7	26.86	36.51	43.57	49.56	54.77	58.81

Table 4.3 PSNR Obtained for "Luigi" by Quantizing Shifted Distance Information Using USQ

Number of Bits		$D^{(2)}$ 2	3	4	5	6	7
$D^{(1)}$	2	20.43	25.45	26.86	27.06	27.09	27.10
	3	25.48	32.22	35.66	36.45	36.63	36.68
	4	26.84	35.62	40.47	42.73	43.45	43.66
	5	27.02	36.35	42.70	46.77	48.85	49.57
	6	27.05	36.52	43.40	48.82	52.85	54.93
	7	27.06	36.57	43.61	49.55	54.89	58.88

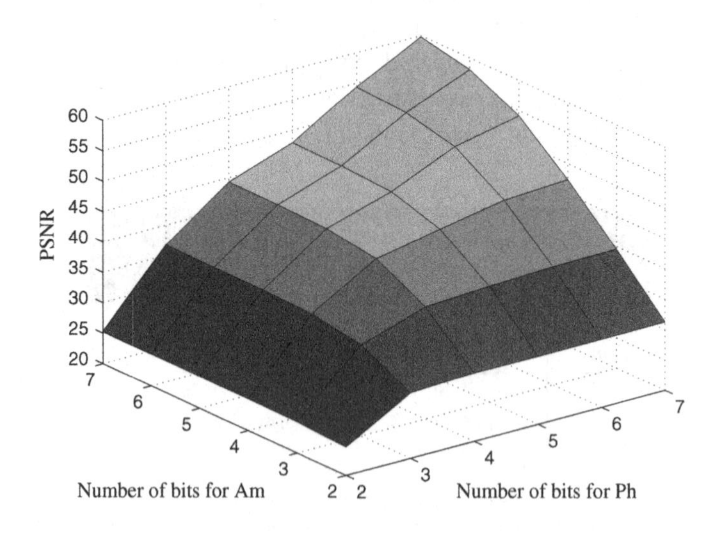

Figure 4.6 Figure corresponding to Table 4.1.

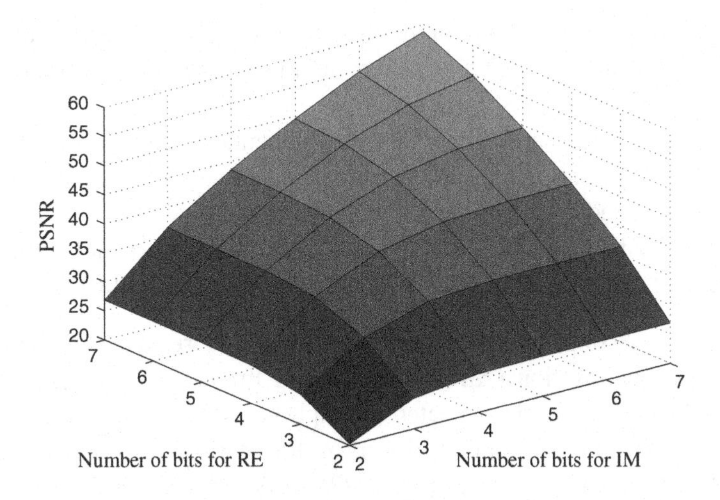

Figure 4.7 Figure corresponding to Table 4.2.

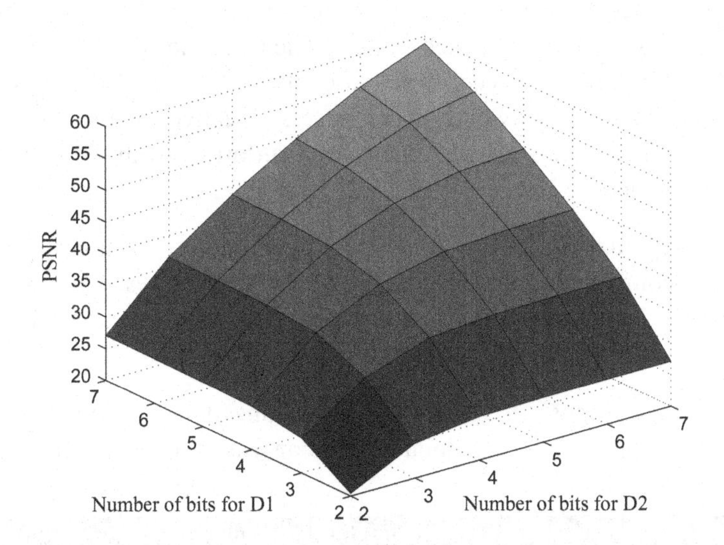

Figure 4.8 Figure corresponding to Table 4.3.

- From Fig. 4.6, the different contributions of amplitude and phase components can easily be observed. At a given bit level for the amplitude component, the change of PSNR value is sharper than that at the same bit level for the phase component. Similarly, it is implied in the numbers in Table 4.1 that the phase component contributes more to the reconstruction than the amplitude component.

This different behavior of the amplitude and phase components is not unexpected, given their obviously different distributions as shown in Fig. 4.5.

- For the real and imaginary components, a quite balanced contribution can be observed from Fig. 4.7. This conclusion is also confirmed by the numerical evaluation in Table 4.2.
- Similarly, for the shifted distance components, a quite balanced contribution can be observed from Fig. 4.8 and Table 4.3.

In the second set of experiments, USQ is replaced by adaptive scalar quantization (ASQ). Considering the probability distributions in Fig. 4.5, ASQ is expected to improve performance further when the distributions are nonuniform. In the experiments, the Lloyd-Max algorithm (Max, 1960; Lloyd, 1982) is adopted for ASQ.

Table 4.4 compares the PSNR obtained from ASQ and USQ on the amplitude-phase and shifted distance information of three objects, with same bit level on each sample. The two kinds of representations similarly affect the reconstruction quality when each sample has the same bit level. However, ASQ is less effective on the amplitude-phase information due to the nearly uniform distribution of phase information.

Redundancy analysis In the third set of experiments, we investigate the inter-component redundancies for the different representations. From Figs. 4.2–4.4, we have a first hint that the real-imaginary and shifted distance information exhibit similar visual patterns.

For this purpose, 2D inter vector quantization is applied to each representation, where an input vector consists of one element from one

Table 4.4 PSNR Obtained From ASQ and USQ on Amplitude-Phase (Am, Ph) and Shifted Distance ($D^{(1)}, D^{(2)}$)

Number of Bits	Luigi				Girl				Bunny			
	Am, Ph		$D^{(1)}, D^{(2)}$		Am, Ph		$D^{(1)}, D^{(2)}$		Am, Ph		$D^{(1)}, D^{(2)}$	
	ASQ	USQ	ASQ	USQ	ASQ	USQ	ASQ	USQ	ASQ	USQ	ASQ	USQ
2	24.74	24.66	25.63	20.43	25.82	26.01	26.71	19.71	19.36	18.77	22.81	15.65
3	35.12	35.06	34.61	32.22	35.73	35.67	34.53	31.63	28.69	28.28	29.71	25.38
4	42.83	42.71	42.44	40.47	43.61	43.32	42.51	40.36	35.26	34.83	36.23	32.33
5	47.27	47.29	48.77	46.77	48.16	48.09	49.35	47.07	41.11	40.82	41.85	38.51
6	53.82	53.80	53.71	52.85	54.72	54.72	54.02	53.04	46.91	46.82	45.73	44.51
7	59.80	59.80	58.89	58.88	60.54	60.57	59.15	59.08	52.90	52.90	50.53	50.52

Table 4.5 PSNR Obtained From LBG-VQ on Amplitude-Phase (Am, Ph) and Shifted Distance Information of Three Objects

	Luigi		Girl		Bunny	
Number of Bits	Am, Ph	$D^{(1)}, D^{(2)}$	Am, Ph	$D^{(1)}, D^{(2)}$	Am, Ph	$D^{(1)}, D^{(2)}$
1	16.15	19.54	16.40	16.99	13.92	16.05
2	17.97	27.61	27.64	29.56	22.39	23.38
3	34.10	35.75	36.36	37.36	29.17	30.52
4	40.62	43.18	42.90	45.15	35.46	37.54
5	46.72	50.26	49.12	52.47	41.58	44.07

component and the corresponding element from the other component. Linde-Buzo-Gray algorithm (LBG-VQ) (Linde et al., 1980) is used in the experiments. Before applying LBG-VQ, it is necessary to normalize the amplitude and phase information, due to the huge difference in their range of values. In the encoding process, the codebook is produced by training all vectors until the MSE is lower than a threshold $\tau = 0.001$. Table 4.5 evaluates the performance of LBG-VQ on amplitude-phase and shifted distance information of three objects. The average bit level varies from 1 to 5.

Reconstructed images for the object "Luigi" obtained by LBG-VQ applied on amplitude-phase and shifted distance representations are given in the left and right column of Fig. 4.9, respectively, at bit level 2, 3, and 4.

The results are summarized as follows:

- It is verified that with increasing codebook size, shifted distance information outperforms amplitude-phase information.
- Comparing with the evaluations of ASQ (see Table 4.4), the impact of applying LBG-VQ on amplitude-phase information is rather limited with gains below 1 dB. In contrast, LBG-VQ applied on shifted distance information leads to significant improvements with gains 3 dB. In other words, higher inter-component redundancies existing in shifted distance information are experimentally confirmed.
- The better performance of shifted distance information is also verified by the visual difference of reconstructed intensities, as shown in Fig. 4.9. Moreover, different from the distortions that could appear in natural images coded by VQ, the intensity of reconstructed objects from quantized holographic information shows surface roughness at high compression.

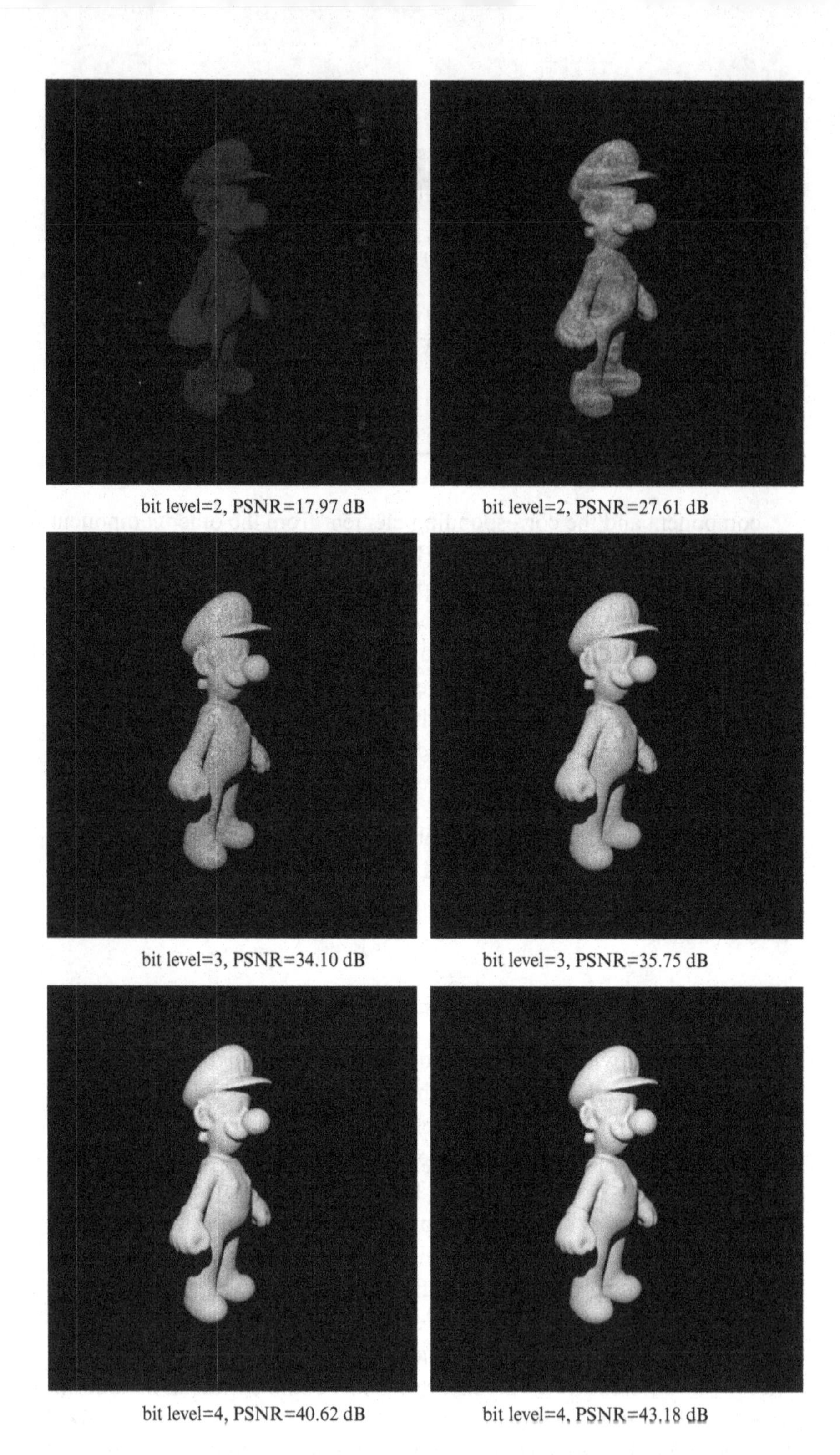

Figure 4.9 Reconstructed images at bit levels of 2, 3, and 4 from amplitude-phase information (left column) and shifted distance information (right column) of "Luigi" object.

CHAPTER 5

Compression of Digital Holographic Data

Even though holography was invented more than 60 years ago, it has received more attention recently thanks to tremendous achievements in computer and digital technologies. Numerous research works have resulted in significant progresses to generate high-resolution digital holograms with cost-effective solutions. However, compression of digital holographic data is still a relatively new field of research.

Figure 5.1 shows the overall schematic of a complete digital holographic communication system. The different components include processing steps for digital hologram acquisition, compression, transmission, decompression, and display.

In detail, a digital hologram is first captured from a still 3D object by using a charged coupled device. For moving 3D objects, a temporal sequence of holograms is generated. Alternatively, the hologram can also be obtained by computational methods from virtual 3D object models. Next, the interference pattern should be converted into a suitable representation. The resulting

Digital Holographic Data Representation and Compression. http://dx.doi.org/10.1016/B978-0-12-802854-4.00005-7

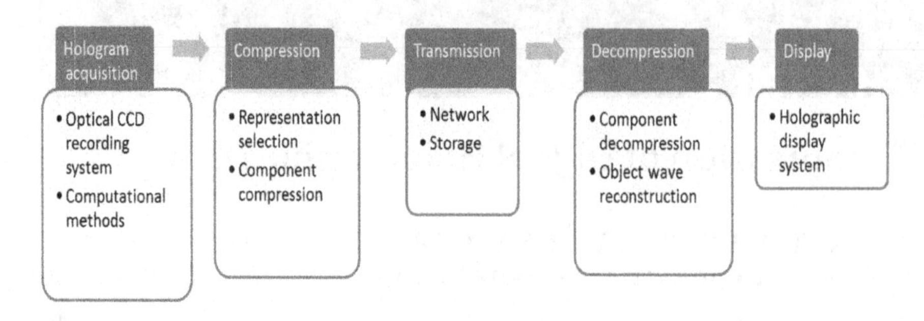

Figure 5.1 Digital holographic communication system components.

representation is then compressed and encoded with properly designed coding tools. The encoded data are then transmitted. At the receiver side, the transmitted is first decoded. This allows reconstruction of the object wave, which can then be rendered on the holographic display device.

5.1 OVERVIEW OF STATE OF THE ART FOR DIGITAL HOLOGRAPHIC DATA COMPRESSION

Both lossless and lossy compression methods have been attempted in compressing digital holographic data. In the early phase, the performance of lossless coding and quantization methods were mainly investigated, while recent research works are more focused on lossy compression with wavelets transform. As summarized in the following, most of the research works on compressing holographic data are categorized into two kinds of approaches: quantization-based and transform-based methods.

5.1.1 Quantization Based

Some lossless and lossy data compression methods are applied to real-imaginary information obtained by phase-shifting interferometry (PSI) in Naughton et al. (2002). Lossless techniques such as Lempel-Ziv, Lempel-Ziv-Welch, Huffman, and Burrows-Wheeler are used to separately code real and imaginary components of the holographic data. However, due to the low spatial redundancies resulting from the speckle nature of holograms, lossless compression methods are usually inefficient (Darakis et al., 2006). As a consequence, lossy compression of holographic data seems essential.

Lossy compression techniques such as subsampling and quantization are applied as well in Naughton et al. (2002). Hologram resampling resulted in a high degradation in reconstructed image quality, but for resizing to a side length of 0.5, a compression rate of 18.6 could be achieved with a high degree of median filtering. Quantization proved to be a very effective technique. Each real and imaginary component could be reduced to 4 bits with an acceptable reconstruction error, resulting in a compression rate of 16.

Mills and Yamaguchi (2005) also confirmed the effectiveness of quantization in both numerical simulation and optical experiments. The influence of bit-depth reduction has been investigated by quantizing the spot-array patterns (the intensity information). They also pointed out that the use of 4 bits appears to be adequate for visual recognition. However, as the number of bits decreases, the resulting quality falls rapidly when only applying quantization.

Furthermore, Naughton et al. (2003) improved the quantization of real-imaginary information with a bit packing operation for real-time networking application. A speedup metric is also defined, which combines space gains due to compression with temporal overheads due to the compression routine and the transmission serialization. It has been reported that quantization to 4 bits results in a compression rate of 16, with normalized root mean square (NRMS) errors in the reconstructed object intensity lower than 0.06.

A companding histogram approach is applied for nonuniform quantization in Shortt et al. (2006b). A companding quantizer is a nonuniform quantizer composed of a compressor, a uniform quantizer, and an expander. A pattern-determined grid is used for the compander, which is fixed-interval sampled. Its pattern is firstly determined by quantization experiments with some DHs in order to produce a hard-coded pattern of clusters. In this way, the computation burden of iterations can be significantly reduced by considering the cluster pattern as a codebook. The compressor and expander works depending on the density of input data. If the input data are dense, the grid is compressed; contrarily, if the input data are sparse, the grid is expanded. The companding approach combines the efficiency of uniform quantization with the improved performance of nonuniform quantization. It exploits *a priori* knowledge of the distribution of the values in the data and performs well when the data distribution can be described. Shortt et al. first applied the companding quantization method for two reasons: (1) the nonuniform distribution of holographic data; and (2) the computation cost and time delay of iterative techniques. They developed two kinds of companding grid: the diamond companding grid and the logarithmic spiral one. The first one is generated based on a logarithmic sampling distribution, which performs as well as some iterative techniques, for example, *k*-means algorithm, with higher efficiency when the number of clusters are in the range of [25, 100]. The second companding grid is developed to improve the performance for smaller numbers of clusters. It uses a logarithmic spiral function to produce the cluster pattern. Improvements have been obtained over the performance of the diamond one.

Similarly, histogram quantization is proposed in Shortt et al. (2007). The histogram quantization technique is a noniterative nonuniform quantization technique, specially designed for digital holographic data with better exploitation of data distribution. In this technique, highest peaks in a representation, for example, real and imaginary data, are extracted to define clusters. The most frequently occurring value in each section is selected as the highest peak. The range r_i of the ith section is defined by

$$r_i = \begin{cases} [a + i\delta, a + (i+1)\delta), & \text{if } i < N-1 \\ [a + i\delta, a + (i+1)\delta], & \text{if } i = N-1, \end{cases} \tag{5.1}$$

where the range of the set of values is $[a, b]$, N is the number of sections, $\delta = \dfrac{(b-a)}{N}$ and $i \in \{0, 1, \ldots, N-1\}$. The peaks from real values and imaginary values are then paired to give complex-valued clusters. Each pixel is quantized to the value of its nearest cluster. The significant advantage of this quantization technique is its less time-complexity under the same requirement of reconstruction quality, compared to iterative techniques. This quantization technique can also be considered as joint quantization, similar to VQ, as the two sets of holographic data are jointly quantized. Shortt et al. also applied this technique on amplitude-phase representation; however, better reconstruction performance is reported for real-imaginary representation, in agreement with the conclusion obtained above.

Differently, based on SQ, a multiple description coding (MDC) method is applied on amplitude-phase information using maximum-a-posteriori in Arrifano et al. (2013). It takes advantage of MDC for optimally coding data between available channels and mitigate channel errors.

A comparative study of several quantization schemes, USQ, ASQ, and VQ, is presented in Xing et al. (2014b). It is shown that jointly encoding the components can bring more benefits for real-imaginary and shifted distance representations that present higher inter-component redundancies. Conversely, the amplitude-phase representation is better encoded separately due to the larger influence of phase information on the reconstruction quality.

5.1.2 Transform Based

Transform-based coding is widely used for image compression, thanks to its ability to compact the signal energy efficiently. Therefore, researchers have naturally investigated the use of transform-based coding, and more specifically wavelet transforms, for the compression of holograms.

Shortt et al. (2006a) introduced wavelet analysis for the compression of real-imaginary components. A 1D-DWT is applied to each component with different resolution levels. The wavelet coefficients are then quantized. It has been found that three levels performed the best on average. In addition, compression of optically obtained and computer-generated phase-shifting interference patterns by standard JPEG and JPEG 2000 compression techniques are, respectively, addressed in Darakis and Soraghan (2006b), with reported compression ratios in the range 20-27 at acceptable reconstruction levels, and Xing et al. (2013a) showing the higher performance of JPEG 2000.

However, standard wavelets are typically designed to process piecewise smooth signals, while holograms contain features which are spread out from the objects. Therefore, applying standard wavelets directly to the hologram is not efficient. For this purpose, Liebling et al. (2003) developed a family of wavelet bases—Fresnelets—obtained by applying the Fresnel transform operator to B-splines bases, which is specially tailored to the specificities of DH. The particular suitability of the Fresnel B-splines has been concluded.

Darakis and Soraghan (2006a) introduced the use of Fresnelets into phase-shifting digital holography (PSDH) for holographic data compression. The real-imaginary components are first separately decomposed to Fresnelet coefficients to a required scale depth. The real and imaginary Fresnelet coefficients are then fed to the SPIHT algorithm. Experimental results verified its extensive flexibility for the compression of PSI holographic data.

However, Viswanathan et al. (2013) analyzed that Fresnelets have limitations for showing localization in frequency regarding a viewpoint-based degraded reconstruction. Instead, they proposed to use Gabor wavelets, which are suitable for measuring the local spatial frequencies so that the coefficients can be pruned corresponding to a viewpoint selection. Also, the angular spectrum method is used to perform the reconstruction instead of the Fresnel transform. The experimental results prove that Gabor wavelets are able to suppress the unwanted orders created in the reconstruction for off-axis holograms and have better time-frequency localization for view-dependent compression techniques. Moreover, they proposed to use Morlet wavelets for transforming a hologram and partly reconstructing a scene by using a sparse set of Morlet transformed coefficients in Viswanathan et al. (2014). It has been shown that a view-dependent representation together with Morlet wavelets form a good starting step for compressing holographic data for next generation 3DTV applications.

On the other hand, Blinder et al. (2013) also investigated wavelet coding on off-axis holograms. Differently, the properties of the off-axis holograms are first examined by independent component analysis, which reveals the importance of orientation and high frequencies in off-axis holograms. For this reason, standard decomposition schemes are not suitable for compressing holograms. Based on the standard JPEG 2000 algorithm, some alternative decomposition schemes are proposed to decompose further the high-frequency subbands. They are combined with direction-adaptive wavelets (Chang and Girod, 2007). Significant improvements have been reported for lossy compression compared with the standard DWT using the Mallat decomposition. Furthermore, they proposed a wavelet packet decomposition scheme combined with directional wavelet transforms for compressing microscopic off-axis holograms in Blinder et al. (2014). Still, the JPEG 2000 standard is modified to cope with holograms by applying the new wavelet decomposition with directional wavelet transforms. This extended JPEG 2000 algorithm shows higher compression efficiency.

Wavelet-based methods are considered effective for compressing holographic data. However, all the coding schemes described above encode different holographic data independently. It could be more efficient to combine the wavelet transform and joint encoding methods to compress phase-shifting holographic data. Xing et al. (2014a) first introduced a joint coding method based on the concept of vector lifting scheme (VLS) to compress shifted distance information. Experimental results indicate the advantage of applying this joint coding scheme. A significant gain of about 2 dB and 0.15 in terms of PSNR and structural similarity (SSIM), respectively, has been achieved compared to independent JPEG 2000 coding scheme. Subsequently, the scheme has been further improved by changing the decomposition structure from a separable to a nonseparable scheme in Xing et al. (2015), resulting in significant improvements.

Several wavelet-based holographic data compression methods are further detailed in Section 5.2.

5.1.3 Extension to Digital Holographic Sequences

Some works on compressing hologram sequences have also been reported. However, most methods are simply based on 2D video compression techniques.

Seo et al. (2006) used a multiview prediction technique and a temporal motion prediction technique to remove the spatial and temporal data redundancies. The predicted and compensated data are then compressed by an MPEG-2 encoder. Enhanced performances have been reported.

The use of MPEG-4 advanced simple profile (ASP) (Pereira and Ebrahimi, 2002) for the compression of hologram sequences was investigated by Darakis and Naughton (2009). Although it has been designed for conventional video, the scheme is effective and achieves good reconstruction quality. Inter-frame coding is also shown to outperform intra-frame coding. In other words, the scheme successfully exploits temporal redundancies in the hologram sequences.

A 3D scanning method is introduced in Seo et al. (2007). More specifically, interference patterns are divided into blocks, and 2D discrete cosine transform (DCT) is performed. The resulting segments are then scanned in 3D, in order to form a video sequence. Finally, the resulting video sequence is encoded using the advanced video coding (AVC) standard—H.264/AVC (Wiegand et al., 2003). The authors show the effectiveness of the scheme and claim significantly improved compression performance when compared to earlier works.

Two video coding schemes, namely H.264/AVC (Wiegand et al., 2003) and the wavelet-based (Dirac, 2009), are compared in Darakis et al. (2010). The authors performed subjective experiments in order to identify the threshold of visually lossless quality. More specifically, the observers are successively shown two sequences in random order, the uncompressed reference and a compressed version, and are asked to identify the compressed one. The bitrate of the compressed sequence is varied until the just noticeable difference threshold is found. While performance is content-dependent, it is shown that compression ratios up 7.5 can be achieved with visually lossless quality.

The recent high efficiency video coding (HEVC) state-of-the-art video coding standard (Sullivan et al., 2012) has also been investigated to compress hologram sequences. In Xing et al. (2013b), HEVC is applied to holograms obtained from animated virtual objects using computer-generated phase-shifting holography. HEVC is shown to achieve high reconstruction quality with a bit rate of 15 Mbps. Moreover, HEVC is demonstrated to be significantly superior to H.264/AVC. Subjective quality assessment experiments using HEVC are also reported in Ahar et al. (2015).

5.2 WAVELET-BASED HOLOGRAPHIC DATA COMPRESSION METHODS

In this section, we discuss in more details several wavelet-based holographic data compression methods based on Fresnelet (Liebling et al., 2003; Darakis and Soraghan, 2006a), separable and nonseparable vector lifting schemes (Xing et al., 2014a, 2015), arbitrary packet decomposition and directional wavelet transform (Blinder et al., 2013, 2014), and Morlet transform (Viswanathan et al., 2014).

5.2.1 Fresnelet Scheme

Fresnelets are a family of B-spline wavelet basis functions, which are specially designed for processing digital holograms (Liebling et al., 2003). Wavelets are efficient for piecewise smooth signals, such as natural images. However, holograms have very different characteristics, and in this case, wavelets do not perform optimally. Fresnelets have been developed in order to address this drawback. They have been successfully applied to compress off-axis and phase-shifting holographic data (Darakis and Soraghan, 2006a). In this part, the basic principles of the design of Fresnelets will be represented.

As explained in Chapter 2, the Fresnel transform of the complex object wave $U(x,y)$ in the wave traveling direction $z = 0$ to the distance $z = d$ can be expressed by 2D convolution integral as

$$\begin{aligned}\hat{U}_\tau(\xi,\eta) &= \frac{\exp(\mathrm{i}kd)}{\mathrm{i}\lambda d}\int_{-\infty}^{+\infty}\int_{-\infty}^{+\infty} U(x,y) \\ &\quad \times \exp\left[\frac{\mathrm{i}\pi}{\lambda d}((\xi-x)^2+(\eta-y)^2)\right]\,\mathrm{d}x\,\mathrm{d}y \\ &= -\mathrm{i}\exp(\mathrm{i}kd)(U * K_\tau)(\xi,\eta),\end{aligned} \tag{5.2}$$

with

$$
\begin{aligned}
K_\tau(x,y) &= \frac{1}{\tau^2}\exp\left[\frac{i\pi}{\tau^2}(x^2+y^2)\right] \\
&= k_\tau(x)\cdot k_\tau(y), \\
\tau &= \sqrt{\lambda d},
\end{aligned}
\tag{5.3}
$$

where $*$ denotes convolution and $K_\tau(x,y)$ is the separable kernel.

The Fresnelet bases are the Fresnel transform of B-splines. B-splines are selected to generate a multiresolution analysis of L_2 because they satisfy the requirements of a valid scaling function of L_2. One-dimensional B-splines of degree n are defined as the $(n+1)$-fold convolution of a rectangular pulse (Unser, 1999):

$$
\beta^n(x) = \underbrace{\beta^0 * \cdots * \beta^0(x)}_{n+1 \text{ times}}, \tag{5.4}
$$

where

$$
\beta^0(x) = \begin{cases} 1, & \text{if } -\frac{1}{2} < x < \frac{1}{2} \\ \frac{1}{2}, & \text{if } |x| = \frac{1}{2} \\ 0, & \text{otherwise.} \end{cases} \tag{5.5}
$$

An alternative equivalent definition of the B-splines is:

$$
\beta^n(x) = \Delta^{n+1} * \frac{(x)_+^n}{n!}, \tag{5.6}
$$

where Δ^{n+1} is the $(n+1)$th centered finite-difference operator:

$$
\Delta^{n+1} = \sum_{k=0}^{n+1} (-1)^k C_{n+1}^k \delta\left(x + \frac{n+1}{2} - k\right) \tag{5.7}
$$

and $(x)_+^n = \max(0,x)^n$ is the one-sided power function. So the explicit central B-splines can be represented by

$$
\beta^n(x) = \sum_{k=0}^{n+1} (-1)^k C_{n+1}^k \frac{\left(x + \frac{n+1}{2} - k\right)_+^n}{n!}. \tag{5.8}
$$

More specially, a two-scale relation of the form can be shown as

$$
\beta^n\left(\frac{x}{2}\right) = \sum_k h(k)\beta^n(x-k), \tag{5.9}
$$

where $h(k)$ is the binomial filter:

$$h(k) = \frac{1}{2^n} C_{n+1}^k. \tag{5.10}$$

Moreover, in Unser (1999), a general family of semi-orthogonal spline wavelets in a two-scale relation has been shown of the form:

$$\psi^n\left(\frac{x}{2}\right) = \sum_k g(k)\beta^n(x-k), \tag{5.11}$$

so a Riesz basis of L_2 is formed by the functions

$$\left\{ \psi_{i,k}^n = 2^{\frac{-i}{2}} \psi(2^{-i}x - k) \right\}. \tag{5.12}$$

These wavelets are linear combinations of B-splines which are specified by $g(k)$, while $g(k)$ is the quadrature mirror filter of $h(k)$, which forms Riesz bases of L_2.

Fresnelets are then obtained by Fresnel transformed B-spline bases:

$$\hat{\beta}_\tau^n(x) = (\beta^n * k_\tau)(x). \tag{5.13}$$

They can be used to decompose the signals into different scales. Particularly, for the two-scale relation, they become

$$\hat{\beta}_{\frac{\tau}{2}}^n\left(\frac{x}{2}\right) = \sum_k h(k)\hat{\beta}^n(x-k). \tag{5.14}$$

The semi-orthogonal ones corresponding to Eq. (5.11) can be derived in the same way.

In order to apply Fresnelets in 2D data, it can be simply derived by extending B-splines to 2D as

$$\beta^n(x,y) = \beta^n(x) \cdot \beta^n(y), \tag{5.15}$$

where $\cdot$ denotes the tensor product. The Fresnelets coefficients of the complex wave $\hat{U}_{\tau_0}$ at scale $j = 1$ can be obtained by

$$\begin{aligned}
\mathbf{LL}\left(\frac{x}{2},\frac{y}{2}\right) &= \hat{U}_{\tau_0} * \left[\hat{\beta}^n_{\tau_0}\left(\frac{x}{2}\right)\cdot\hat{\beta}^n_{\tau_0}\left(\frac{y}{2}\right)\right]\\
\mathbf{HL}\left(\frac{x}{2},\frac{y}{2}\right) &= \hat{U}_{\tau_0} * \left[\hat{\psi}^n_{\tau_0}\left(\frac{x}{2}\right)\cdot\hat{\beta}^n_{\tau_0}\left(\frac{y}{2}\right)\right]\\
\mathbf{LH}\left(\frac{x}{2},\frac{y}{2}\right) &= \hat{U}_{\tau_0} * \left[\hat{\beta}^n_{\tau_0}\left(\frac{x}{2}\right)\cdot\hat{\psi}^n_{\tau_0}\left(\frac{y}{2}\right)\right]\\
\mathbf{HH}\left(\frac{x}{2},\frac{y}{2}\right) &= \hat{U}_{\tau_0} * \left[\hat{\psi}^n_{\tau_0}\left(\frac{x}{2}\right)\cdot\hat{\psi}^n_{\tau_0}\left(\frac{y}{2}\right)\right].
\end{aligned} \tag{5.16}$$

For more scales decomposition, the subband **LL** needs to be decomposed further.

In Darakis and Soraghan (2006a), the obtained Fresnelets coefficients obtained from real-imaginary information are further quantized by uniform quantization and compressed by SPIHT coding. The comparison is conducted between lossless B-spline wavelets transform coded hologram and SPIHT coded Fresnelet coefficients. For the same compression rate, the Fresnelets-based scheme leads to smaller NRMS.

5.2.2 Separable Vector Lifting Scheme

As shown in Chapter 4, in the cases of shifted distance and real-imaginary representations, the two components of the hologram show similar visual content. Thus, it is expected that these kinds of data include redundancies, and from this point of view, efficient hologram compression schemes can be designed by exploiting the dependencies between these patterns. For this purpose, Xing et al. (2014a, 2015) first proposed to apply a joint coding method, based on the concept of a vector lifting scheme (VLS), to phase-shifting holographic data.

Most of the existing joint coding schemes, which have been developed in the literature for video and stereo/multiview data compression purpose, consist of two steps. Assuming that two input correlated images are available to be encoded, the first step consists of selecting one image as a reference and encoding it independently of the other one. Then, the second image, selected as a target image, is predicted from the first one, and the difference between the two images, called the residual, is encoded. Typically, DCT or DWT can be used for encoding both the reference and residual images (Moellenhoff and Maier, 1998; Boulgouris and Strintzis, 2002). Contrary to this standard scheme, the main feature of VLS is that it does not generate a residual image,

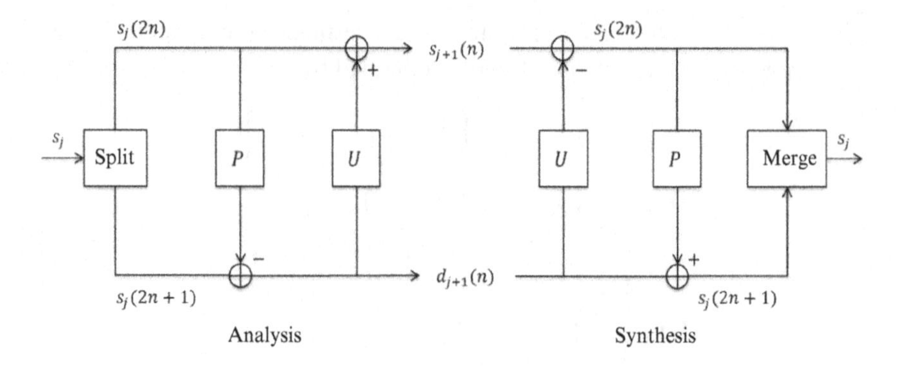

Figure 5.2 Generic lifting structure.

but two compact representations of both images. VLS is an extended scheme based on lifting scheme (LS) (Sweldens, 1996; Hampson and Pesquet, 1998; Kaaniche et al., 2011a, 2012). LS is an alternative method for computing DWT, which is simpler and faster than the classical one. The original signal can also be easily obtained by the inverse transform. With these advantages, LS has been proven to be an efficient tool for still image coding. Many extensions have been applied to LS. One of them known as the quincunx lifting scheme has been found particularly useful for coding satellite images by using a quincunx grid (Gouze et al., 2004). Moreover, some directional transforms, such as oriented wavelet transform (Chappelier and Guillemot, 2006) and grouplet (Mallat, 2009), have also been developed with LS for image processing.

Figure 5.2 shows a generic separable LS structure for 1D signal with one level resolution. Biorthogonal wavelets can be constructed by a series of operators including *split*, *predict*, and *update* in the forward transform (analysis step).

Split First, the input 1D signal s_j is partitioned into two sample sets: the even samples $s_j(2n)$ and the odd samples $s_j(2n+1)$.

Predict Next, predict each sample in one of the subsets by the neighboring samples in the other subset. A prediction error or detail signal, for the case of predicting odd sample (the same for the following contents), can be computed:

$$d_{j+1}(n) = s_j(2n+1) - \mathbf{p}^\top \mathbf{s}_j(n), \tag{5.17}$$

where $\mathbf{p}$ is the prediction vector containing the prediction weights, $\mathbf{s}_j(n) = (s_j(2n - 2k))_{k\in\mathcal{P}}$ is the reference vector containing the even samples for predicting the odd sample, and $\mathcal{P}$ is the support of the predictor.

Update Then, a coarser approximation of the original signal is generated by a smoothing operation on the even samples using the detail coefficients:

$$s_{j+1}(n) = s_j(2n) + \mathbf{u}^{\top}\mathbf{d}_{j+1}(n), \tag{5.18}$$

where similarly, $\mathbf{u}$ is the update vector containing the weights of update operator, $\mathbf{d}_{j+1}(n) = (d_{j+1}(n-k))_{k\in\mathcal{U}}$ is the reference vector containing the detail coefficients, and $\mathcal{U}$ is the support of the update operator.

It is obvious that the two sample subsets can be easily reconstructed from the forward transform by two inverse operators:

Undo update

$$s_j(2n) = s_{j+1}(n) - \mathbf{u}^{\top}\mathbf{d}_{j+1}(n), \tag{5.19}$$

Undo predict

$$s_j(2n+1) = d_{j+1}(n) + \mathbf{p}^{\top}\mathbf{s}_j(n). \tag{5.20}$$

Then, they are merged to obtain the original signal. Above all, the LS is easier to implement than the filter bank-based method.

Similarly, the separable 2D-DWT can also be implemented by applying the 1D case to the lines and columns. Also, one can obtain a multiresolution representation by recursively repeating the analysis steps to the approximation coefficients.

Based on LS, VLS has been developed to encode jointly two sets of dependent signals which will be denoted by $S^{(1)}$ and $S^{(2)}$ (Benazza-Benyahia et al., 2002; Kaaniche et al., 2009). The block diagram of a separable vector lifting scheme is shown in Fig. 5.3.

The principle of this multiscale decomposition is described for a given line x. In what follows, $S_j^{(1)}$ and $S_j^{(2)}$ designate the approximation coefficients of 2D signals, $S^{(1)}$ and $S^{(2)}$, at resolution level j. While $j = 0$ corresponds

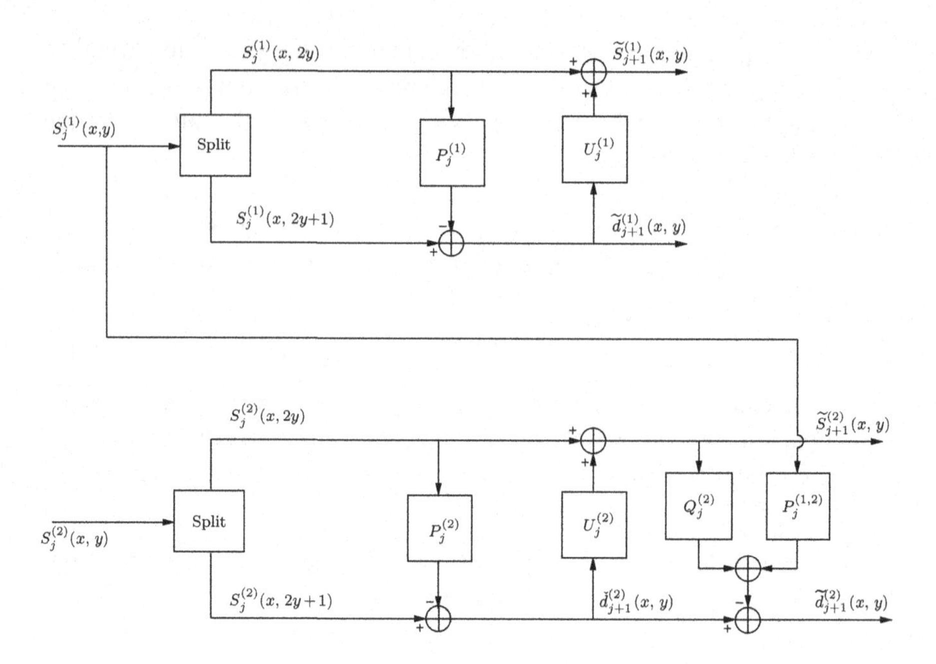

Figure 5.3 Principle of the VLS decomposition.

to the initial (full resolution) signal $S^{(1)}$ and $S^{(2)}$, note that the dimensions of $S_j^{(1)}$ and $S_j^{(2)}$ $(j \geq 1)$ are divided by 2^j along the horizontal and vertical directions.

Decomposition of reference signal As can be seen in Fig. 5.3, the reference signal $S^{(1)}$ is firstly encoded using a classical lifting structure composed of a prediction and an update stage. To this end, for a given line x, the input signal $S_j^{(1)}(x, y)$ is firstly partitioned into two datasets formed by the even $S_j^{(1)}(x, 2y)$ and odd samples $S_j^{(1)}(x, 2y+1)$, respectively. Then, during the prediction step, each sample of one of the two subsets (say the odd ones) is predicted from the neighboring even samples, yielding the detail coefficients $\tilde{d}_{j+1}^{(1)}$ at the resolution $(j+1)$:

$$\tilde{d}_{j+1}^{(1)}(x, y) = S_j^{(1)}(x, 2y+1) - \sum_{k \in \mathcal{P}_j^{(1)}} p_{j,k}^{(1)} S_j^{(1)}(x, 2y-2k), \qquad (5.21)$$

where the coefficients $p_{j,k}^{(1)}$ and the set $\mathcal{P}_j^{(1)}$ represent, respectively, the weights and the support of the predictor of the odd samples

$S_j^{(1)}(x, 2y+1)$. After that, the update step aims at computing a coarser approximation $\tilde{S}_{j+1}^{(1)}$ of the original signal by smoothing the even sample $S_j^{(1)}(x, 2y)$ as follows:

$$\tilde{S}_{j+1}^{(1)}(x,y) = S_j^{(1)}(x,2y) + \sum_{k \in \mathcal{U}_1^{(1)}} u_{j,k}^{(1)} \tilde{d}_{j+1}^{(1)}(x, y-k), \tag{5.22}$$

where the set $\mathcal{U}_j^{(1)}$ denotes the spatial support of the update operator whose coefficients are $u_{j,k}^{(1)}$.

Decomposition of target signal Once the reference signal $S^{(1)}$ is encoded in intra mode, the attention will be paid now to the target signal $S^{(2)}$. It is important to note that the main difference between a basic lifting scheme and the vector lifting scheme is that for the target signal $S^{(2)}$, the prediction step uses samples from itself and *also* their corresponding samples taken from the reference signal $S^{(1)}$.

As shown in Fig. 5.3, a P-U-P structure is used for the target signal $S^{(2)}$. More precisely, a first intra-prediction step is applied to generate an intermediate detail signal $\check{d}_{j+1}^{(2)}$, which serves to compute the approximation signal $\tilde{S}_{j+1}^{(2)}$ through the update step. After that, an hybrid prediction is performed by exploiting *simultaneously* the intra- and inter-redundancies in order to compute the final detail signal $\tilde{d}_{j+1}^{(2)}$. Thus, the resulting decomposition implies the following equations:

$$\check{d}_{j+1}^{(2)}(x,y) = S_j^{(2)}(x,2y+1) - \sum_{k \in \mathcal{P}_j^{(2)}} p_{j,k}^{(2)} S_j^{(2)}(x, 2y-2k), \tag{5.23}$$

$$\tilde{S}_{j+1}^{(2)}(x,y) = S_j^{(2)}(x,2y) + \sum_{k \in \mathcal{U}_j^{(2)}} u_{j,k}^{(2)} \check{d}_{j+1}^{(2)}(x, y-k), \tag{5.24}$$

$$\tilde{d}_{j+1}^{(2)}(x,y) = \check{d}_{j+1}^{(2)}(x,y) - \left(\sum_{k \in \mathcal{Q}_j} q_{j,k} \tilde{S}_{j+1}^{(2)}(x, y-k) + \sum_{k \in \mathcal{P}_j^{(1,2)}} p_{j,k}^{(1,2)} S_j^{(1)}(x, 2y+1-k) \right), \tag{5.25}$$

where $\mathcal{P}_j^{(2)}$ (respectively, $\mathcal{P}_j^{(1,2)}$) is the spatial support of the intra-signal (respectively, inter-signals) whereas its weights are designated by $p_{j,k}^{(2)}$ (respectively, $p_{j,k}^{(1,2)}$), and $\mathcal{Q}_j$ (respectively, $q_{j,k}$) is the support (respectively, weights) of the second intra-signal predictor.

Since a separable decomposition has been considered, these steps are iterated on the columns y of the resulting subbands $\tilde{S}_{j+1}^{(1)}$, $\tilde{d}_{j+1}^{(1)}$, $\tilde{S}_{j+1}^{(2)}$, $\tilde{d}_{j+1}^{(2)}$ in order to produce the approximation subbands $S_{j+1}^{(1)}$ and $S_{j+1}^{(2)}$ as well as three details subbands, for each signal, oriented horizontally, vertically, and diagonally. This decomposition is again repeated on the approximation subbands over J resolution levels, yielding the multiresolution representation of the two input signals.

Finally, at the last resolution level J, instead of encoding the approximation subband of the target signal $S_J^{(2)}$, it is proposed to encode the residual subband given by:

$$e_J^{(2)}(x,y) = S_J^{(2)}(x,y) - \sum_{k \in \mathcal{P}_J^{(1,2)}} p_{J,k}^{(1,2)} S_J^{(1)}(x, y-k). \tag{5.26}$$

Prediction and update filters Prediction and update filters with spatial supports $\mathcal{P}_j^{(1)} = \{-1, 0\}$ and $\mathcal{U}_j^{(1)} = \{0, 1\}$ are considered for signal $S^{(1)}$. In addition, the weights of the update filter are set for all the resolution levels to: $u_{j,0}^{(1)} = u_{j,1}^{(1)} = \frac{1}{4}$. However, it is proposed to optimize the prediction weights $p_{j,-1}^{(1)}$ and $p_{j,0}^{(1)}$ in order to design a coding scheme well adapted to the content of the hologram data. Since the detail coefficients can be viewed as prediction errors, the prediction filter coefficients can be optimized at each resolution level j by minimizing the variance of the detail signal $\tilde{d}_{j+1}^{(1)}$. More specifically, the well-known Yule-Walker equations are applied to solve the optimization problem:

$$\mathbf{E}\left[\mathbf{S}_j(x,y)\mathbf{S}_j(x,y)^\top\right]\mathbf{p}_j = \mathbf{E}\left[S_j^{(1)}(x, 2y+1)\mathbf{S}_j(x,y)\right], \tag{5.27}$$

where

$$\mathbf{S}_j(x,y) = \left(S_j^{(1)}(x, 2y), S_j^{(1)}(x, 2y+2)\right)^\top. \tag{5.28}$$

$\mathbf{p}_j = (p_{j,0}^{(1)}, p_{j,-1}^{(1)})^\top$ is the prediction weighting vector and $\mathbf{E}[\cdot]$ denotes the mathematical expectation.

Concerning the second signal $S^{(2)}$, the same intra-prediction and update filters used with $S^{(1)}$ will be employed to generate the signals $\check{d}^{(2)}_{j+1}$ and $\tilde{S}^{(2)}_{j+1}$. Then, the second prediction stage is performed by setting $\mathcal{Q}_j = \{-1, 0\}$ and $\mathcal{P}^{(1,2)}_j = \{-1, 0, 1\}$ for $j \in \{0, \ldots, J-1\}$ and $\mathcal{P}^{(1,2)}_J = \{0\}$. The coefficients $q_{j,k}$ and $p^{(1,2)}_{j,k}$ are also optimized by minimizing the variance of the detail signal $\tilde{d}^{(2)}_{j+1}$ by Yule-Walker equations.

The subband coefficients resulting from the VLS decomposition are then quantized and encoded using the EBCOT algorithm similarly to the JPEG 2000 standard (see Chapter 4).

The effectiveness of the joint coding scheme based on separable VLS for holographic data compression purposes has been validated in Xing et al. (2014a). More specifically, the authors compared the performance of the VLS scheme with two reference methods using the shifted distance holographic data representation. The first method corresponds to the state-of-the-art hologram compression technique where the inputs $D^{(1)}$ and $D^{(2)}$ are separately encoded by using existing still image coders. To this end, the 9/7 wavelet transform retained in the lossy compression mode of JPEG 2000 has been used. In what follows, this scheme will be designated by "Independent." The second one is the standard joint coding scheme, where the reference image $D^{(1)}$ and a residual one, given by $D^{(2)} - D^{(1)}$, are also encoded by applying the 9/7 transform. We recall that this technique has been considered in most of joint coding schemes developed in the context of stereo and video data compression. It will be designated by "Standard." Note that JPEG 2000 has been used as an entropy encoder for these different hologram compression methods. Figure 5.4 shows the corresponding rate-distortion results in terms of PSNR, respectively, SSIM, versus the bitrate given in bit per pixel (bpp). Both PSNR and SSIM are computed between the original and reconstructed objects. For more details about the experimental results, readers can refer to Xing et al. (2014a).

5.2.3 Nonseparable Vector Lifting Scheme

The VLS approach presented in Section 5.2.2 has been performed in a separable way by cascading the 1D decomposition along the horizontal direction, then along the vertical direction. However, according to the visual

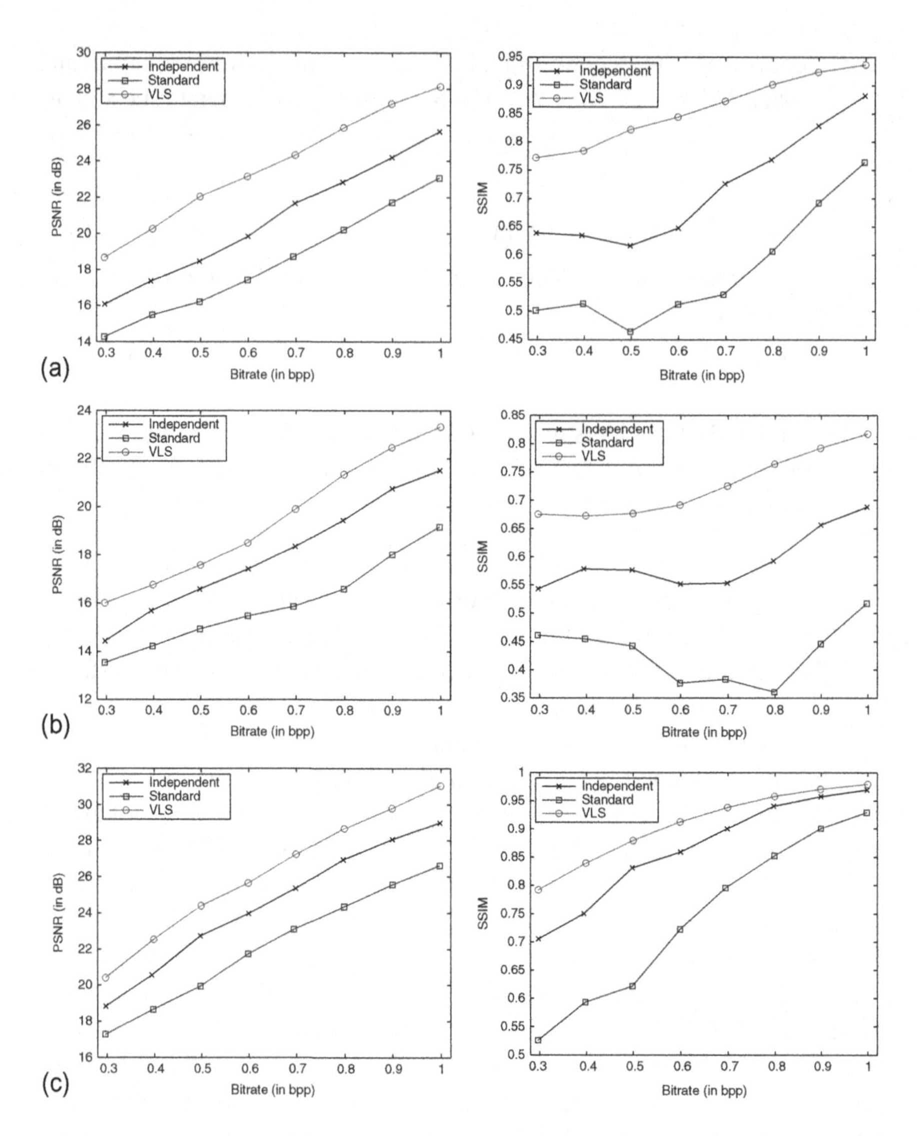

Figure 5.4 Rate-distortion performance of three different hologram compression schemes: independent, standard, and VLS, applied on $D^{(1)}$ *and* $D^{(2)}$*, for the objects: (a) "Bunny-1," (b) "Bunny-2," and (c) "Girl."*

patterns of the difference data $D^{(1)}$ and $D^{(2)}$, it can be noticed that these signals present some structures that are neither horizontal nor vertical.

Therefore, to exploit better the characteristics of the hologram data, a nonseparable decomposition has been proposed in Xing et al. (2015), with the objective to build an efficient content-adaptive decomposition. This 2D nonseparable vector lifting scheme (NS-VLS) is described in more detail in this section. We use the same notations as in Section 5.2.2.

5.2.3.1 Basic Concept of Decomposition Structure

At each pixel location (x,y), the approximation coefficients of the first (respectively, second) image $D_j^{(1)}(x,y)$ (respectively, $D_j^{(2)}(x,y)$) are divided into four polyphase components denoted by

$$\begin{cases} D_{0,j}^{(i)}(x,y) = D_j^{(i)}(2x,2y), \\ D_{1,j}^{(i)}(x,y) = D_j^{(i)}(2x,2y+1), \\ D_{2,j}^{(i)}(x,y) = D_j^{(i)}(2x+1,2y), \\ D_{3,j}^{(i)}(x,y) = D_j^{(i)}(2x+1,2y+1), \end{cases} \tag{5.29}$$

where $i \in \{1,2\}$.

Figure 5.5 shows the NS-VLS analysis structure. One image (here $D^{(1)}$) is selected as a reference image and encoded independently of the other

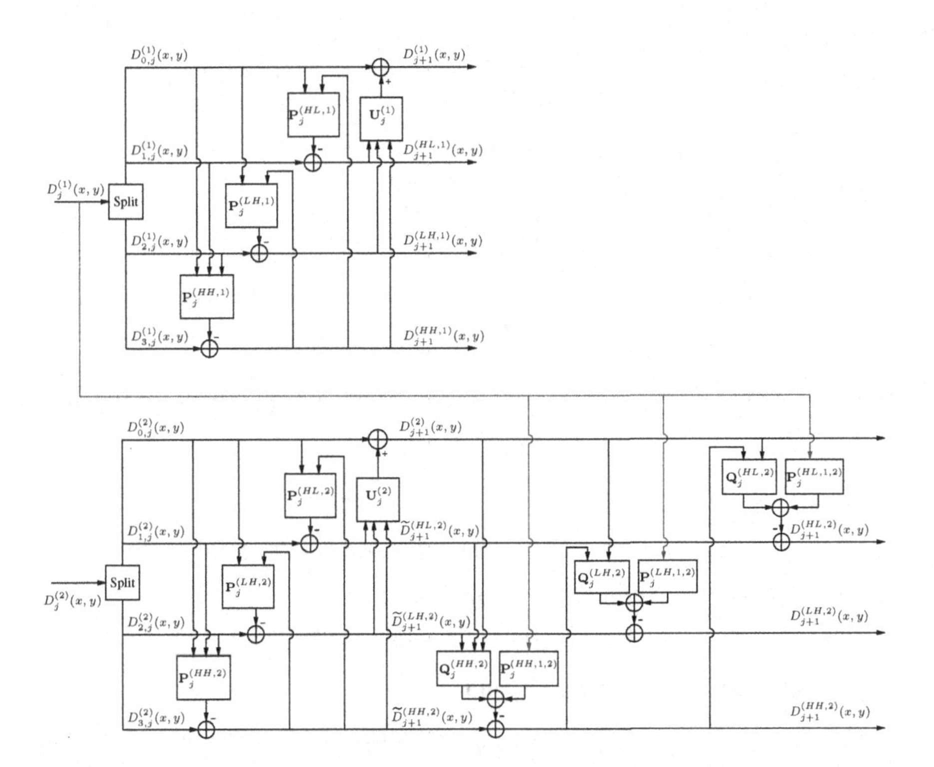

Figure 5.5 Proposed NS-VLS decomposition structure.

one. To this end, a classical nonseparable lifting structure (Kaaniche et al., 2011b), composed of three prediction steps and an update one, is applied to $D_j^{(1)}$ in order to generate the detail signals oriented diagonally $D_{j+1}^{(\mathrm{HH},1)}$, vertically $D_{j+1}^{(\mathrm{LH},1)}$, and horizontally $D_{j+1}^{(\mathrm{HL},1)}$, as well as the approximation one $D_{j+1}^{(1)}$. Thus, the output wavelet coefficients, at the resolution level $(j+1)$, can be written as follows:

$$D_{j+1}^{(\mathrm{HH},1)}(x,y) = D_{3,j}^{(1)}(x,y) - \left(\left(\mathbf{P}_{0,j}^{(\mathrm{HH},1)}\right)^{\top}\mathbf{D}_{0,j}^{(\mathrm{HH},1)} + \left(\mathbf{P}_{1,j}^{(\mathrm{HH},1)}\right)^{\top}\mathbf{D}_{1,j}^{(\mathrm{HH},1)} + \left(\mathbf{P}_{2,j}^{(\mathrm{HH},1)}\right)^{\top}\mathbf{D}_{2,j}^{(\mathrm{HH},1)}\right), \tag{5.30}$$

$$D_{j+1}^{(\mathrm{LH},1)}(x,y) = D_{2,j}^{(1)}(x,y) - \left(\left(\mathbf{P}_{0,j}^{(\mathrm{LH},1)}\right)^{\top}\mathbf{D}_{0,j}^{(\mathrm{LH},1)} + \left(\mathbf{P}_{1,j}^{(\mathrm{LH},1)}\right)^{\top}\underline{\mathbf{D}}_{j+1}^{(\mathrm{HH},1)}\right), \tag{5.31}$$

$$D_{j+1}^{(\mathrm{HL},1)}(x,y) = D_{1,j}^{(1)}(x,y) - \left(\left(\mathbf{P}_{0,j}^{(\mathrm{HL},1)}\right)^{\top}\mathbf{D}_{0,j}^{(\mathrm{HL},1)} + \left(\mathbf{P}_{1,j}^{(\mathrm{HL},1)}\right)^{\top}\overline{\mathbf{D}}_{j+1}^{(\mathrm{HH},1)}\right), \tag{5.32}$$

$$D_{j+1}^{(1)}(x,y) = D_{0,j}^{(1)}(x,y) + \left(\left(\mathbf{U}_{0,j}^{(\mathrm{HL},1)}\right)^{\top}\mathbf{D}_{j+1}^{(\mathrm{HL},1)} + \left(\mathbf{U}_{1,j}^{(\mathrm{LH},1)}\right)^{\top}\mathbf{D}_{j+1}^{(\mathrm{LH},1)} + \left(\mathbf{U}_{2,j}^{(\mathrm{HH},1)}\right)^{\top}\mathbf{D}_{j+1}^{(\mathrm{HH},1)}\right), \tag{5.33}$$

where for every $i \in \{0,1,2\}$ and $o \in \{\mathrm{HL}, \mathrm{LH}, \mathrm{HH}\}$,

- $\mathbf{P}_{i,j}^{(o,1)} = (p_{i,j}^{(o,1)}(s,t))_{(s,t)\in\mathcal{P}_{i,j}^{(o,1)}}$ represents the prediction weighting vector whose support is denoted by $\mathcal{P}_{i,j}^{(o,1)}$;
- $\mathbf{D}_{i,j}^{(o,1)} = (D_{i,j}^{(1)}(x+s,y+t))_{(s,t)\in\mathcal{P}_{i,j}^{(o,1)}}$ is a reference vector used to compute $D_{j+1}^{(o,1)}(x,y)$;
- $\underline{\mathbf{D}}_{j+1}^{(\mathrm{HH},1)} = (D_{j+1}^{(\mathrm{HH},1)}(x+s,y+t))_{(s,t)\in\mathcal{P}_{1,j}^{(\mathrm{LH},1)}}$ and $\overline{\mathbf{D}}_{j+1}^{(\mathrm{HH},1)} = (D_{j+1}^{(\mathrm{HH},1)}(x+s,y+t))_{(s,t)\in\mathcal{P}_{1,j}^{(\mathrm{HL},1)}}$ correspond to two reference vectors used in the second and the third prediction steps, respectively;
- $\mathbf{U}_{i,j}^{(o,1)} = (u_{i,j}^{(o,1)}(s,t))_{(s,t)\in\mathcal{U}_{i,j}^{(o,1)}}$ is the update vector coefficients whose support is designated by $\mathcal{U}_{i,j}^{(o,1)}$; and
- $\mathbf{D}_{j+1}^{(o,1)} = (D_{j+1}^{(o,1)}(x+s,y+t))_{(s,t)\in\mathcal{U}_{i,j}^{(o,1)}}$ is the reference vector containing the set of detail samples used in the update step.

For the second image $D_j^{(2)}$, selected as a target one, a joint wavelet decomposition is performed by taking into account its correlation with the reference one $D_j^{(1)}$. More specifically, a lifting structure, similar to that used with $D_j^{(1)}$, is firstly applied on $D_j^{(2)}$ in order to produce three intermediate detail signals $\tilde{D}_{j+1}^{(\mathrm{HH},2)}$, $\tilde{D}_{j+1}^{(\mathrm{LH},2)}$, and $\tilde{D}_{j+1}^{(\mathrm{HL},2)}$, which will be used to compute the approximation coefficients $D_{j+1}^{(2)}$. Then, a second prediction stage is added.

Indeed, three hybrid prediction steps, which aim at exploiting simultaneously the intra- and inter-image redundancies, are applied in order to generate the final detail coefficients $D_{j+1}^{(\mathrm{HH},2)}$, $D_{j+1}^{(\mathrm{LH},2)}$, and $D_{j+1}^{(\mathrm{HL},2)}$. Thus, the final detail wavelet coefficients, at the resolution level $(j+1)$, are expressed as follows:

$$
\begin{aligned}
D_{j+1}^{(\mathrm{HH},2)}(x,y) =& \tilde{D}_{j+1}^{(\mathrm{HH},2)}(x,y) - \Bigg(\left(\mathbf{Q}_{0,j}^{(\mathrm{HH},2)}\right)^{\top} \tilde{\mathbf{D}}_{0,j+1}^{(\mathrm{HH},2)} \\
&+ \left(\mathbf{Q}_{1,j}^{(\mathrm{HH},2)}\right)^{\top} \tilde{\mathbf{D}}_{1,j+1}^{(\mathrm{HH},2)} + \left(\mathbf{Q}_{2,j}^{(\mathrm{HH},2)}\right)^{\top} \tilde{\mathbf{D}}_{2,j+1}^{(\mathrm{HH},2)} \\
&+ \left(\mathbf{P}_{0,j}^{(\mathrm{HH},1,2)}\right)^{\top} \mathbf{D}_{0,j}^{(\mathrm{HH},1)} + \left(\mathbf{P}_{1,j}^{(\mathrm{HH},1,2)}\right)^{\top} \mathbf{D}_{1,j}^{(\mathrm{HH},1)} \\
&+ \left(\mathbf{P}_{2,j}^{(\mathrm{HH},1,2)}\right)^{\top} \mathbf{D}_{2,j}^{(\mathrm{HH},1)} + \left(\mathbf{P}_{3,j}^{(\mathrm{HH},1,2)}\right)^{\top} \mathbf{D}_{3,j}^{(\mathrm{HH},1)}\Bigg),
\end{aligned}
$$

$$
\begin{aligned}
D_{j+1}^{(\mathrm{LH},2)}(x,y) =& \tilde{D}_{j+1}^{(\mathrm{LH},2)}(x,y) - \Bigg(\left(\mathbf{Q}_{0,j}^{(\mathrm{LH},2)}\right)^{\top} \tilde{\mathbf{D}}_{0,j+1}^{(\mathrm{LH},2)} \\
&+ \left(\mathbf{Q}_{1,j}^{(\mathrm{LH},2)}\right)^{\top} \underline{\mathbf{D}}_{j+1}^{(\mathrm{HH},2)} + \left(\mathbf{P}_{0,j}^{(\mathrm{LH},1,2)}\right)^{\top} \mathbf{D}_{0,j}^{(\mathrm{LH},1)} \\
&+ \left(\mathbf{P}_{2,j}^{(\mathrm{LH},1,2)}\right)^{\top} \mathbf{D}_{2,j}^{(\mathrm{LH},1)}\Bigg),
\end{aligned}
$$

$$
\begin{aligned}
D_{j+1}^{(\mathrm{HL},2)}(x,y) =& \tilde{D}_{j+1}^{(\mathrm{HL},2)}(x,y) - \Bigg(\left(\mathbf{Q}_{0,j}^{(\mathrm{HL},2)}\right)^{\top} \tilde{\mathbf{D}}_{0,j+1}^{(\mathrm{HL},2)} \\
&+ \left(\mathbf{Q}_{1,j}^{(\mathrm{HL},2)}\right)^{\top} \overline{\mathbf{D}}_{j+1}^{(\mathrm{HH},2)} + \left(\mathbf{P}_{0,j}^{(\mathrm{HL},1,2)}\right)^{\top} \mathbf{D}_{0,j}^{(\mathrm{HL},1)} \\
&+ \left(\mathbf{P}_{1,j}^{(\mathrm{HL},1,2)}\right)^{\top} \mathbf{D}_{1,j}^{(\mathrm{HL},1)}\Bigg),
\end{aligned}
$$

where for every $i \in \{0, 1, 2, 3\}$ and $o \in \{\mathrm{HL}, \mathrm{LH}, \mathrm{HH}\}$,

- $\mathbf{Q}_{i,j}^{(o,2)} = (q_{i,j}^{(o,2)}(s,t))_{(s,t)\in\mathcal{Q}_{i,j}^{(o,2)}}$ is the intra-prediction weighting vector whose support is designated by $\mathcal{Q}_{i,j}^{(o,2)}$;
- $\mathbf{P}_{i,j}^{(o,1,2)} = (p_{i,j}^{(o,1,2)}(s,t))_{(s,t)\in\mathcal{P}_{i,j}^{(o,1,2)}}$ is the hybrid prediction weighting vector whose support is denoted by $\mathcal{P}_{i,j}^{(o,1,2)}$;
- $\tilde{\mathbf{D}}_{0,j+1}^{(o,2)} = (D_{j+1}^{(2)}(x+s,y+t))_{(s,t)\in\mathcal{Q}_{0,j}^{(o,2)}}$ is a reference vector containing the approximation coefficients $D_{j+1}^{(2)}$ used to compute the detail ones $D_{j+1}^{(o,2)}(x,y)$;
- $\tilde{\mathbf{D}}_{1,j+1}^{(\mathrm{HH},2)} = (\tilde{D}_{j+1}^{(\mathrm{HL},2)}(x+s,y+t))_{(s,t)\in\mathcal{Q}_{1,j}^{(\mathrm{HH},2)}}$ and $\tilde{\mathbf{D}}_{2,j+1}^{(\mathrm{HH},2)} = (\tilde{D}_{j+1}^{(\mathrm{LH},2)}(x+s,y+t))_{(s,t)\in\mathcal{Q}_{2,j}^{(\mathrm{HH},2)}}$ are two reference vectors, containing, respectively, the intermediate detail coefficients $\tilde{D}_{j+1}^{(\mathrm{HL},2)}$ and $\tilde{D}_{j+1}^{(\mathrm{LH},2)}$, used to compute the final detail coefficients $D_{j+1}^{(\mathrm{HH},2)}(x,y)$;
- $\underline{\mathbf{D}}_{j+1}^{(\mathrm{HH},2)} = (D_{j+1}^{(\mathrm{HH},2)}(x+s,y+t))_{(s,t)\in\mathcal{Q}_{1,j}^{(\mathrm{LH},2)}}$ and $\overline{\mathbf{D}}_{j+1}^{(\mathrm{HH},2)} = (D_{j+1}^{(\mathrm{HH},2)}(x+s,y+t))_{(s,t)\in\mathcal{Q}_{1,j}^{(\mathrm{HL},2)}}$ are two intra-prediction vectors used to compute $D_{j+1}^{(\mathrm{LH},2)}(x,y)$ and $D_{j+1}^{(\mathrm{HL},2)}(x,y)$; and
- $\mathbf{D}_{i,j}^{(o,1)} = (D_{i,j}^{(1)}(x+s,y+t))_{(s,t)\in\mathcal{P}_{i,j}^{(o,1,2)}}$ is a vector containing some samples of the reference image $D_j^{(1)}$ used to exploit the inter-image redundancies during the computation of the final detail coefficients $D_{j+1}^{(o,2)}(x,y)$.

By repeating the same decomposition strategy on the approximation subbands over J resolution levels, two multiresolution representations of $D^{(1)}$ and $D^{(2)}$ are obtained. Finally, at the last resolution level J, instead of encoding the approximation subband of the target image $D_J^{(2)}$, it would be interesting to exploit its correlation with $D_J^{(1)}$ and thus to encode the following residual subband $e_J^{(2)}$:

$$e_J^{(2)}(x,y) = D_J^{(2)}(x,y) - p_J^{(1,2)} D_J^{(1)}(x,y), \tag{5.34}$$

where $p_J^{(1,2)}$ is a hybrid prediction coefficient that exploits the correlation between $D_J^{(2)}$ and $D_J^{(1)}$.

5.2.3.2 Optimization Techniques

For the same reason as the VLS-based approach, the ultimate aim is to design an NS-VLS-based decomposition well adapted to the characteristics of the holographic data. Consequently, the optimization of all the operators is necessary. Due to the complexity of the NS-VLS structure, different strategies of optimization are applied:

Optimization of the predictors for $D^{(1)}$ For the first image $D^{(1)}$, the different prediction filters $\mathbf{P}_j^{(o,1)}$ (with $o \in \{\text{HH, LH, HL}\}$), used to generate the detail subbands $D_{j+1}^{(o,1)}$, are optimized at each resolution level by minimizing the variance of the detail coefficients. As mentioned before, the Yule-Walker equations should be satisfied.

Optimization of the update operators Different from the fixed assignment of the update operators in separable VLS, the update filter here at each level is optimized by minimizing the error between the approximation coefficients $D_{j+1}^{(1)}$ and the decimated version of the signal resulting from an ideal low-pass filter. Due to the complexity of the equations for the application, the details will not be provided here. The readers can refer to Kaaniche et al. (2011b) for more details.

Optimization of the operators for $D^{(2)}$ For the second image $D^{(2)}$, all the prediction and update operators can also be optimized by adopting the same strategy used with $D^{(1)}$. However, since $D^{(1)}$ and $D^{(2)}$ present similar content, the optimization process of the first three prediction filters $\mathbf{P}_j^{(o,2)}$ and the update one $\mathbf{U}_j^{(2)}$ can be omitted by imposing these operators to be equal to those obtained with $D^{(1)}$. Thus, we have:

$$\begin{aligned} &\mathbf{P}_j^{(\text{HH},2)} = \mathbf{P}_j^{(\text{HH},1)}, \quad \mathbf{P}_j^{(\text{LH},2)} = \mathbf{P}_j^{(\text{LH},1)}, \\ &\mathbf{P}_j^{(\text{HL},2)} = \mathbf{P}_j^{(\text{HL},1)}, \quad \mathbf{U}_j^{(2)} = \mathbf{U}_j^{(1)}. \end{aligned} \tag{5.35}$$

Note that this procedure presents the advantages of simplifying the optimization strategy and reducing the overhead cost corresponding to the number of filter coefficients that must be sent to the decoder. Finally, for the remaining hybrid prediction filters $\mathbf{P}_j^{(o,1,2)}$ and $\mathbf{Q}_j^{(o,2)}$, they will be optimized by minimizing the variance of the detail coefficients $D_{j+1}^{(o,2)}$.

Figure 5.6 shows the improved performance of NS-VLS, compared to independent coding scheme, standard joint coding scheme, and separable VLS, when using the shifted distance holographic data representation.

In addition, different prediction filter lengths L_p have also been investigated. Indeed, increasing L_p may be interesting for two reasons. The first one

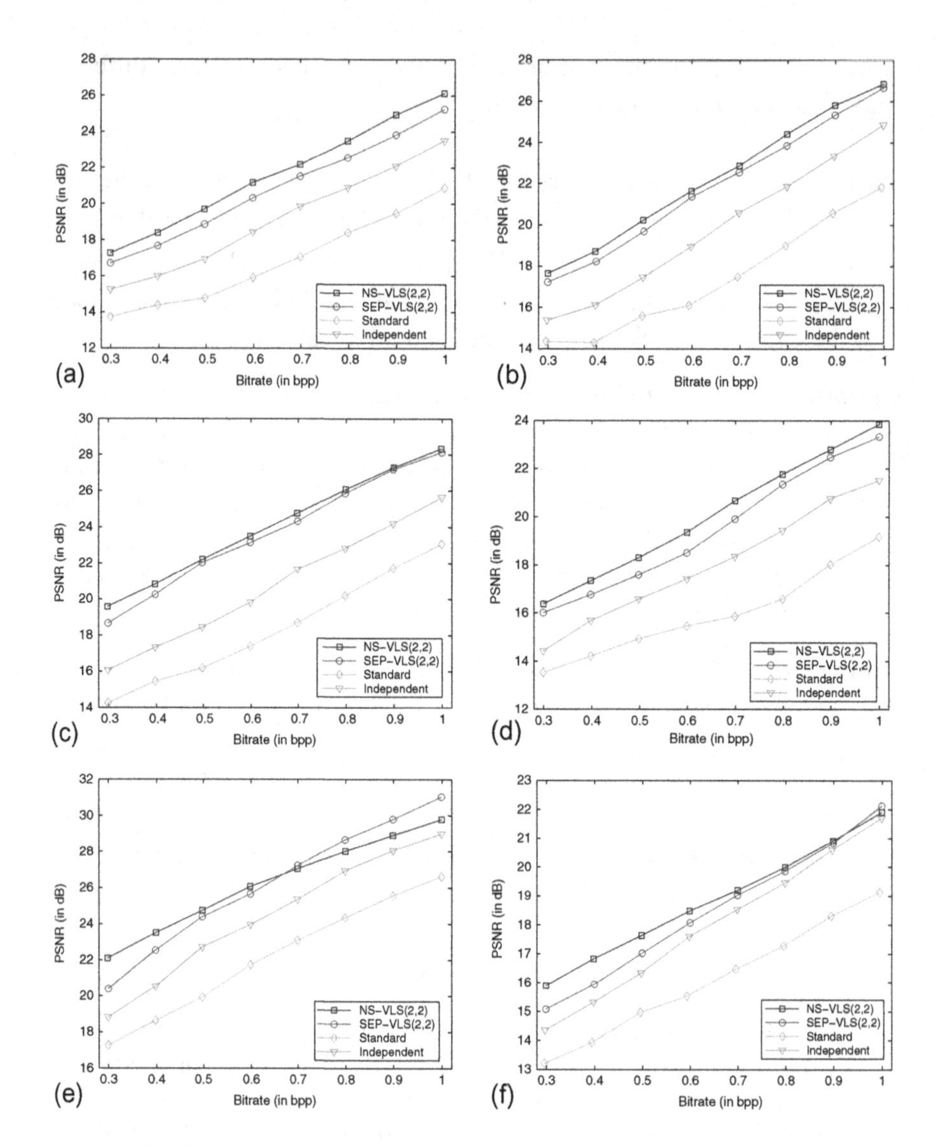

Figure 5.6 Rate-distortion performance of four different hologram compression schemes: independent, standard, separable VLS, and NS-VLS, applied on $D^{(1)}$ and $D^{(2)}$, for the objects: (a) "Luigi-1," (b) "Luigi-2," (c) "Bunny-1," (d) "Bunny-2," (e) "Girl," and (f) "Teapot."

is explained by the fact that the holographic data present repetitive circular structures similar to the propagation of waves. The second one is due to the objective of VLS, which consists of exploiting the inter-images redundancies through the prediction stage. In this respect, in addition to the case given by $L_p = 2$, other cases with $L_p \in \{6, 12, 20, 32\}$ are also considered. The structures are designated by NS-VLS(2, 2), NS-VLS(6, 2), NS-VLS(12, 2),

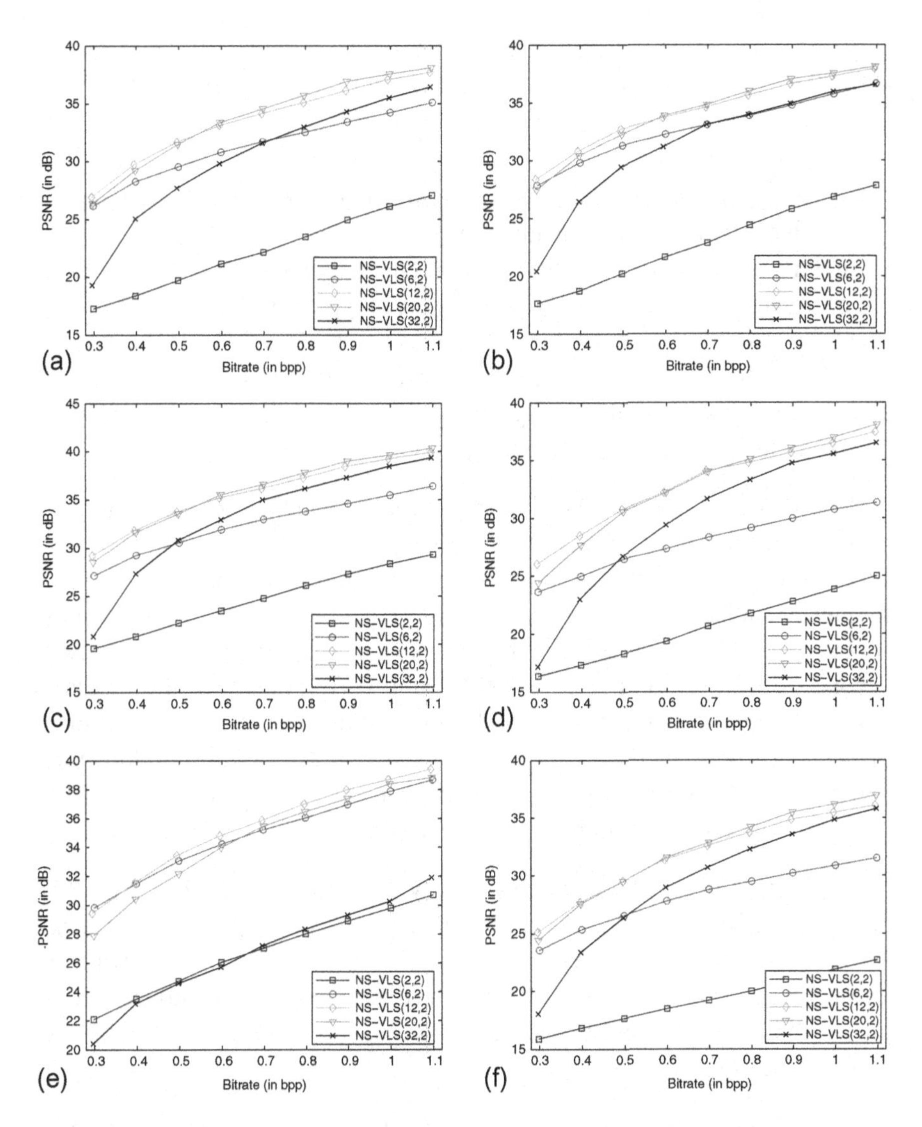

Figure 5.7 Rate-distortion performance of NS-VLS with different prediction filter lengths for the objects: (a) "Luigi-1," (b) "Luigi-2," (c) "Bunny-1," (d) "Bunny-2," (e) "Girl," and (f) "Teapot."

NS-VLS(20, 2), and NS-VLS(32, 2). Some results are given in Fig. 5.7. Compared to the case $L_{\mathrm{p}} = 2$, it can be observed that increasing this length up to 12 or 20 leads to greatly improved compression performance. However, very long filters ($L_{\mathrm{p}} = 32$) lead to worse performance at low bitrate because of the expensive cost of encoding the filter coefficients compared to the case of small L_{p} values. Noted that the best filter length may differ according to the different content of holograms. For more details about the experimental results, readers can refer to Xing et al. (2015).

5.2.4 Arbitrary Packet Decomposition and Direction-Adaptive Discrete Wavelet Transform

Digital holographic data exhibit different characteristics when compared to natural images. For instance, whereas images typically have a $1/f^2$ power spectrum distribution, holographic data are characterized by significantly larger amount of high-frequency content. Moreover, the fringe patterns show a significant directionality. Straightforwardly, these characteristics need to be taken into account when designing a coding scheme.

In order to address the higher spatial frequency content, packet decompositions can be applied. More specifically, in contrast with the Mallat decomposition, packet decompositions further decompose the higher-frequency subbands, in addition to the decomposition of the low-frequency subbands, in order to achieve higher energy compaction. Such packet decompositions are supported in JPEG 2000 Part 2; however, the standard does not support a homogeneous decomposition style when the number of wavelet levels increases. For this purpose, a JPEG 2000 architecture compliant decomposition mechanism is proposed in Blinder et al. (2013, 2014) to fully enable arbitrary packet decompositions.

Exploiting directionality is also clearly an important property. For this purpose, in Blinder et al. (2013, 2014), it is proposed to replace the classical JPEG 2000 wavelet transform by a direction-adaptive discrete wavelet transform (DA-DWT). More precisely, the direction-adaptive wavelet filters introduced by Chang and Girod (2007) are used, in combination with the above wavelet packet decomposition.

The resulting coding architecture is shown in Fig. 5.8, where multiple DA-DWTs are applied on each evenly sized blocks of the hologram.

The DA-DWT is also based on the idea of lifting scheme, to be precise, the directional lifting based. Moreover, it is able to represent more efficiently the sharp features in images. In order to explain DA-DWT, we follow the notations in Chang and Girod (2007). Let $\Pi = \{(l_x, l_y) \in \mathbb{Z}^2\}$ be a 2D orthogonal sampling grid composed of four subgrids: $\Pi_{pq} = \{(l_x, l_y) \in \Pi | l_x \bmod 2 = p, l_y \bmod 2 = q\}$. $\mathbf{S} = \{s[\mathbf{l}]\}$, where $s[\mathbf{l}] = s[l_x, l_y]$ and $\mathbf{l} = (l_x, l_y)$, denotes a set of image samples on Π. $\mathbf{S}_0 = \{s[\mathbf{l}_0] | \mathbf{l}_0 \in$

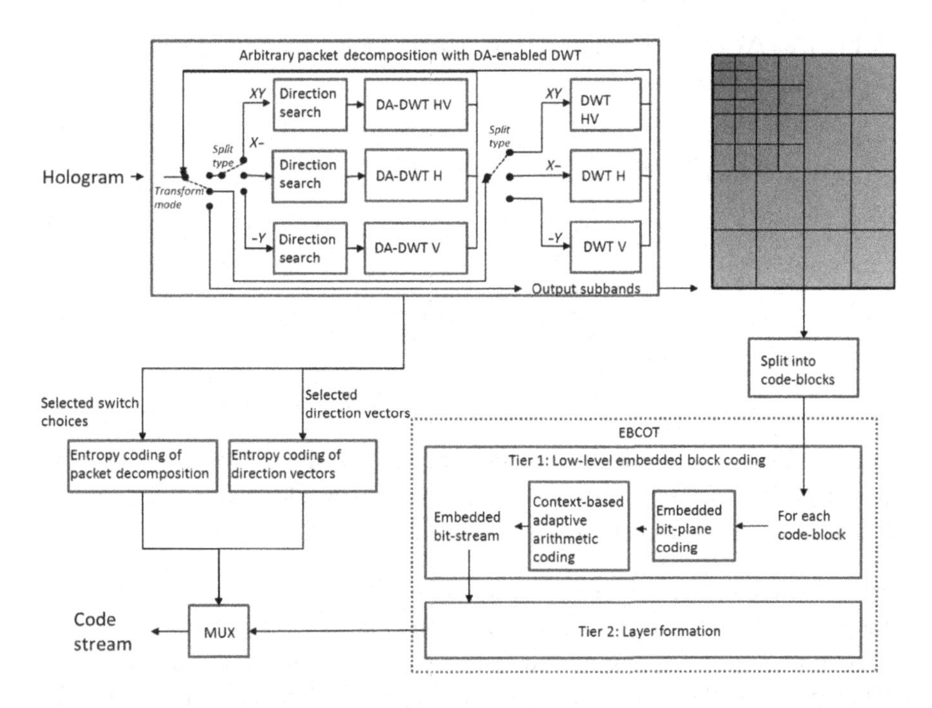

Figure 5.8 JPEG 2000 architecture extended with an arbitrary packet decomposition and direction-adaptive discrete wavelet transform. Source: Adapted from Blinder et al. (2014)

$\Pi_0 = \Pi_{00} \cup \Pi_{01}\}$ and $\mathbf{S}_1 = \{s[\mathbf{l}_1] | \mathbf{l}_1 \in \Pi_1 = \Pi_{10} \cup \Pi_{11}\}$ denote the samples in even and odd rows, respectively.

Similar to the description of the lifting scheme in Section 5.2.2, the detail signal $\mathbf{w}_1 = \{w_1[\mathbf{l}_1], \mathbf{l}_1 \in \Pi_1\}$ and approximation signal $\mathbf{w}_0 = \{w_0[\mathbf{l}_0], \mathbf{l}_0 \in \Pi_0\}$ can be obtained by a prediction step and an update step in 1D vertical lifting of samples in $\mathbf{S}_0$, respectively, as follows:

$$\begin{aligned} w_1[\mathbf{l}_1] &= g_{\mathrm{H}} \cdot (s[\mathbf{l}_1] - \mathbf{P}_{s,\mathbf{l}_1}(\mathbf{S}_0)), \quad \forall \mathbf{l}_1 \in \Pi_1 \\ w_0[\mathbf{l}_0] &= g_{\mathrm{L}} \cdot (s[\mathbf{l}_0] - g_{\mathrm{H}}^{-1} \cdot \mathbf{U}_{s,\mathbf{l}_0}(\mathbf{w}_1)), \quad \forall \mathbf{l}_0 \in \Pi_0, \end{aligned} \tag{5.36}$$

where g_{H}, g_{L} are scaling factors and $\mathbf{P}(\cdot)$, $\mathbf{U}(\cdot)$ are prediction and update operators, respectively. After further decomposition along the columns, four subbands in total are generated: $\mathbf{w}_{00} = \mathbf{LL}$ defined on Π_{00}, $\mathbf{w}_{01} = \mathbf{LH}$ defined on Π_{01}, $\mathbf{w}_{10} = \mathbf{HL}$ defined on Π_{10}, and $\mathbf{w}_{11} = \mathbf{HH}$ defined on Π_{11}.

In DA-DWT, the prediction and update operators with direction $\mathbf{d} = (d_x, d_y)$ are defined as

$$\mathbf{P}^{i}_{s,\mathbf{l}_1}(\mathbf{s}_0) = \sum_{k=-K_P}^{K_P-1} c_{P,k} \cdot [\mathbf{l}_1 - (2k+1)\mathbf{d}]$$
$$\mathbf{U}_{s,\mathbf{l}_0}(\mathbf{w}_1) = \sum_{k=-K_U}^{K_U-1} c_{U,k} \cdot \sum_{\{\mathbf{l}_1 | \mathbf{l}_1-(2k+1)\mathbf{d}^*_{\mathbf{l}_1}=\mathbf{l}_0\}} w_1[\mathbf{l}_1], \tag{5.37}$$

where $\mathbf{d}$ is defined such as to satisfy:

$$\mathbf{l}_1 - (2k+1)\mathbf{d} \in \Pi_0, \quad \forall\, \mathbf{l}_1 \in \Pi_1, \; k = -K_P, \ldots, K_P - 1, \tag{5.38}$$

and $i = 0, 1, \ldots, N_c - 1$ is the direction index; N_c is the number of prediction directions; K_P, K_U, $c_{P,k}$, and $c_{U,k}$ are defined by the adopted wavelet kernel; and $\mathbf{d}^*_{\mathbf{l}_1}$ denotes the direction selected at location $\mathbf{l}_1$ for $\mathbf{P}_{s,\mathbf{l}_1}(\mathbf{s}_0)$.

For each image block, the optimum direction $\mathbf{d}$ is selected by minimizing the prediction error $w_1[\mathbf{l}_1]$. A speed-up process to search for the optimum direction has been also proposed in Stevens et al. (2008). For a multiscale decomposition, the directional lifting steps above are repeated for every **LL** subband.

Due to the high overhead cost for storing the direction indices of each subband, the DA-DWT are only applied for three main decompositions involved in the generation of the 1**LL**, 2**LL**, and 3**LL** subbands (Blinder et al., 2013).

Significant improvements in coding performance have been reported for off-axis holography and computer-generated holography compared to the conventional JPEG 2000 standard, with Bjøntegaard delta-peak signal-to-noise ratio improvements ranging up to 11 dB for lossy compression in the 0.125-2.00 bits per pixel (bpp) range, and bit-rate reductions of up to 1.6 bpp for lossless compression (Blinder et al., 2013, 2015). Table 5.1 demonstrates the significant improvements resulting from the wavelet packet decomposition and DA-DWT on five different test objects.

5.2.5 Morlet Transform

The problem of viewpoint scalability is considered in Viswanathan et al. (2013). The authors address different aspects related to hologram display in the presence of a few viewers. In such a case, only a sparse set of

Table 5.1 BD-PSNR Improvements in (dB) Compared to the DWT Mallat, per Decomposition, With and Without DA-DWT

Lossy	Pack1	Pack2	Full Packet	DA-DWT+ Mallat	DA-DWT+ Pack1	DA-DWT+ Pack2	DA-DWT+ Full Packet
Neuron	3.45	3.34	2.79	5.75	3.40	3.28	2.75
Erythocryte	5.44	5.41	8.20	7.28	8.31	8.30	8.39
Microlenses	0.06	0.30	2.10	2.89	2.65	2.61	2.37
Ball	11.81	11.91	12.64	5.14	12.74	12.74	12.57
Scratch	−7.13	−7.07	−5.26	1.98	−4.59	−4.57	−4.41
Average	2.73	2.78	4.09	4.59	4.50	4.47	4.33

Source: Blinder et al. (2013).

holographic data is necessary to reconstruct a given viewpoint. In other words, the wavelet transform should allow coefficient pruning so that only relevant data are transmitted. To enable this view-based representation of the hologram, good space-frequency localization is needed. Since the wavelet families based on or derived from Gabor basis functions can provide the best space-frequency localization, in Viswanathan et al. (2014), Morlet wavelets are proposed to transform off-axis holograms.

Morlet wavelets are based on the concepts of Gabor functions. In order to design the Morlet wavelets, we start from the concepts of Gabor functions. A Gabor function is a Gaussian kernel multiplied by a sinusoid. Its 1D continuous form is defined as

$$g_{\beta,f_0}(x) = K \cdot \exp(-\beta^2 x^2) \exp(2i\pi f_0 x) \quad \beta^2 = \frac{1}{2\sigma^2}, \tag{5.39}$$

where f_0 is the frequency, K is the norm of the basis function, and σ is the standard deviation of the Gaussian function. In order to have equal number of oscillations for all frequencies, the setting

$$f_0 = \frac{1}{\sqrt{2}} \pi \beta \tag{5.40}$$

is chosen. The 1D Morlet mother wavelet is further defined as

$$\psi(x) = K \cdot \exp(-\beta^2 x^2) \left(\exp(2i\pi f_0 x) - \exp\left(-\frac{\pi^2}{\beta^2} f_0^2 \right) \right), \tag{5.41}$$

where the term $\exp\left(-\frac{\pi^2}{\beta^2} f_0^2 \right)$ is introduced to eliminate the DC term of the off-axis hologram (Zhong et al., 2009). By extending Eq. (5.41) to 2D, we obtain

$$\psi(x,y) = K \cdot \exp\left(-\beta^2\left(x^2+y^2\right)\right)\left(\exp\left(2i\pi\left(\mu_0 x + \nu_0 y\right)\right) - \exp\left(-\frac{\pi^2}{\beta^2}\left(\mu_0^2+\nu_0^2\right)\right)\right), \tag{5.42}$$

where μ_0 and ν_0 are the spatial frequencies in x and y directions, respectively; K is the L_2-norm of the basis. For the same purpose of having equal number of oscillations for all frequencies, the setting

$$\mu_0^2 + \nu_0^2 = \pi^2\beta^2 \tag{5.43}$$

is chosen. Consequently, we can obtain

$$\exp\left(-\frac{\pi^2}{\beta^2}\left(\mu_0^2+\nu_0^2\right)\right) = \exp\left(-\pi^4\right) \approx 0. \tag{5.44}$$

Representing the spatial frequencies μ_0 and ν_0 into polar form

$$\mu_0 = F_0\cos\theta, \quad \nu_0 = F_0\sin\theta, \tag{5.45}$$

where the variation of θ describes the variation of the spatial frequencies in x and y directions. Substituting Eqs. (5.40), (5.43), and (5.45) to Eq. (5.42), the 2D Morlet mother wavelet becomes

$$\psi(x,y) = K \cdot \exp\left(-\frac{2f_0^2\left(x^2+y^2\right)}{\pi^2}\right)\exp\left(2i\pi^2\beta\left(x\cos\theta + y\sin\theta\right)\right), \tag{5.46}$$

where the mother wavelet is centered at f_0. In order to build the family of Morlet wavelets, the scaling parameter s for controlling the spatial and frequency resolution of decomposition is introduced to span the frequency plane, where the scaled frequencies are defined by

$$f = \frac{f_0}{s}. \tag{5.47}$$

Therefore, the continuous 2D Morlet wavelet can be represented with the parameters s and θ by

$$\psi_{s,\theta}(x,y) = K \cdot \exp\left(-\frac{2f_0^2}{s^2\pi^2}\left(x^2+y^2\right)\right)\exp\left(2i\pi^2\frac{f_0}{s\pi}\left(x\cos\theta + y\sin\theta\right)\right). \tag{5.48}$$

For further application, the continuous transform has to be discretized to obtain the view-based representation of holograms. The discretization is mainly conducted from three aspects: the discretization of x and y, the discretization of s, and the discretization of θ.

The discretization of x and y is based on the consideration of the Gaussian function. Assume that the length of the Gaussian is L which is selected based on the tapering of the Gaussian function and there are N_0 oscillations of the sinusoid with frequency f_0, we have

$$L = \frac{N_0}{f_0}. \tag{5.49}$$

By Nyquist criterion, the sampling frequency f_s should be twice of f_0. We consequently have the actual size of the analog window function

$$L_\omega = L \cdot f_s = \frac{N_0}{f_0} \cdot 2f_0 = 2N_0. \tag{5.50}$$

The ration parameter θ is descretized by performing N_θ rotations of the sinusoid from 0 to π. The discrete spatial frequencies f_x and f_y in x and y directions are then represented by

$$\begin{aligned} f_x &= \frac{f_c \cos(\Theta r)}{s}, \\ f_y &= \frac{f_c \sin(\Theta r)}{s}, \end{aligned} \tag{5.51}$$

respectively, where $\Theta = \frac{\pi}{N_\theta}$, r is the discrete rotation parameter and $f_c = \frac{1}{\Delta}$ where Δ is the pixel period of the holographic display.

The discretization of the scale parameter s is dependent on the discrete viewer positions. Assume that θ_x and θ_y are the diffraction angles along x and y directions, respectively. Then the viewer plan can be divided into zones R_{θ_x,θ_y}, each of which corresponds to a unique Morlet transformed hologram zone H_{θ_x,θ_y}. The discrete diffraction angles are obtained by

$$\begin{aligned} \theta_x &= \arcsin\left(\frac{\lambda f_c \cos(\Theta r)}{s}\right), \\ \theta_y &= \arcsin\left(\frac{\lambda f_c \sin(\Theta r)}{s}\right). \end{aligned} \tag{5.52}$$

Above all, the discrete Morlet wavelet can be derived as

$$\psi_{s,r}(m,n) = K \cdot \exp\left(-\frac{f_c^2}{s^2\pi^2}\left(m^2+n^2\right)\right)$$
$$\exp\left(2i\pi^2\frac{f_c}{s\pi}\left(m\cos\left(\Theta r\right)+n\sin\left(\Theta r\right)\right)\right), \quad (5.53)$$

where $-N_0 \leqslant m \leqslant N_0$ and $-N_0 \leqslant n \leqslant N_0$. The Morelet wavelet transform is then performed on the off-axis holograms.

CHAPTER 6

Concluding Remarks

Digital holography offers appealing features for 3D imaging applications, overcoming some inherent limitations of current stereoscopic and multiview technologies, with the potential to become the ultimate 3D experience. However, for this vision to become reality, several major technological hurdles need to be overcome. In particular, digital holography requires a very high data rate. Therefore, a key challenge is to define an effective and compact data representation in order to be able to efficiently handle, store, and transmit digital holograms.

In this book, we have addressed this issue and presented an overview of existing techniques for the compression of holographic data. This topic has gained significant interest in recent years, with various directions explored and several innovative solutions proposed. Most of the existing research works have heavily built upon existing compression techniques developed for still images and video sequences, including JPEG and MPEG suites of standards. Nevertheless, digital holograms exhibit rather different features and characteristics, when compared to natural imagery content. Clearly, better statistical modeling of holograms is needed in order to design tailored compression techniques. It is also important to take into account the multidisciplinary nature of the problem of dealing with challenges in optics, signal processing, and compression.

Despite the existing body of published research, more efforts are needed. In particular, we can identify the following open issues which need further investigations.

First, very different settings have been proposed to generate digital holograms, for instance, on-axis versus off-axis or digital recording versus CGH. Moreover, a number of important parameters come into play, including the hologram dimensions, the pixel size, and the wavelength. It results in a wide parameters space which may strongly impact the final hologram and its characteristics. It is therefore paramount to carry out systematic and

Digital Holographic Data Representation and Compression. http://dx.doi.org/10.1016/B978-0-12-802854-4.00006-9

comprehensive investigations in order to understand better the effect of this wide parameters space. In turn, this also has consequences on the efficiency of subsequent compression techniques.

Second, it is currently impossible to compare results from different publications, as they are using different datasets obtained with different configurations. It is straightforward that common datasets are needed in order to allow researchers to meaningfully compare various approaches. As a first step in this direction, an open database for experimental validations of holographic compression engines was recently proposed in Blinder et al. (2015).[1] It is currently composed of five computer-generated holograms. In addition, it is providing three types of data representation: intensity interference patterns, complex wavefields, and phase-only holograms. However, given the wide parameters space for digital holography as discussed above, further efforts are needed in order to create more comprehensive datasets.

Third, better performance assessment methodologies need to be defined. Fidelity can be measured at two stages: on the compressed hologram or on the reconstructed object, the second way being *a priori* more meaningful. In addition, most previous research works have used objective quality measures, notably PSNR and SSIM, developed and validated for natural still images. Given that objects reconstructed by digital holography exhibit very different characteristics, this methodology remains questionable until it is methodically validated. In parallel, the visual perception related to the viewing of holograms need to be investigated and understood. To the best of our knowledge, the first subjective experiments to assess the impact of digital holography compression have been carried out in Darakis et al. (2010). More recently, comprehensive subjective studies are reported in Ahar et al. (2015). On a closely related topic, Finke et al. (2015) provides insight in visual perception aspects related to holographic displays. These efforts need to be continued in order to develop and validate objective quality measures which are well-suited in the context of digital holography.

Fourth, most published research publications in this domain have considered simplistic scenes composed of one or very few objects. This is in large part due to limitations of current technology, which is only capable of capturing scenes with small angular size. In order to live up to the expectation of digital holography as the ultimate 3D experience,

[1] This open database is available at http://www.erc-interfere.eu/.

it is paramount to consider more complex scenes representative of realistic application scenarios.

Last but not least, another key open issue is to understand at which stage of the processing pipeline compression needs to be performed. Most related publications have considered the compression of holograms, as reported in this book. However, CGH allows for another approach, namely to transmit a conventional representation of the 3D scene and to compute the hologram at the display end. This alternative results in two benefits. Firstly, given that a conventional representation of the 3D scene is transmitted, it can be efficiently compressed with traditional 3D video coding schemes. Hence, the development of new compression techniques specifically for holography data is avoided. Secondly, with such a scheme, the representation of the scene can be completely decoupled from the display device, thus offering greater flexibility. Such an approach is proposed in Senoh et al. (2014) for holographic TV. It is based on multiview video coding and depth map coding, and holograms are generated at the receiving end. The authors claim that the proposed scheme uses 1/97,000 of the data rate when compared to the compression and transmission of holograms, while achieving an equivalent subjective quality.

While significant progresses have already been achieved in recent years, by tackling the above open issues, further significant performance gains can be expected in the coming years, thus opening new horizons for digital holography.

BIBLIOGRAPHY

Abookasis, D., Rosen, J., 2006. Three types of computer-generated hologram synthesized from multiple angular viewpoints of a three-dimensional scene. Appl. Opt. 45 (25), 6533-6538. http://dx.doi.org/10.1364/AO.45.006533. URL http://ao.osa.org/abstract.cfm?URI=ao-45-25-6533.

Ahar, A., Blinder, D., Bruylants, T., Schretter, C., Munteanu, A., Schelkens, P., 2015. Subjective quality assessment of numerically reconstructed compressed holograms. In: Proc. SPIE, Applications of Digital Image Processing XXXVIII, San Diego, CA, USA.

Ahrenberg, L., Benzie, P., Magnor, M., Watson, J., 2008. Computer generated holograms from three dimensional meshes using an analytic light transport model. Appl. Opt. 47 (10), 1567-1574. http://dx.doi.org/10.1364/AO.47.001567. URL http://ao.osa.org/abstract.cfm?URI=ao-47-10-1567.

Arrifano, A., Antonini, M., Pereira, M., 2013. Multiple description coding of digital holograms using maximum-a-posteriori. In: European Workshop on Visual Information Processing, pp. 232-237.

Benazza-Benyahia, A., Pesquet, J.C., Hamdi, M., 2002. Vector lifting schemes for lossless coding and progressive archival of multispectral images. IEEE Trans. Geosci. Remote Sens. 40 (9), 2011-2024.

Bildkompression, E.W.-B., der Technik, S., 2003. Embedded wavelet-based image compression: state of the art. it Inf. Technol. 45, 5.

Blinder, D., Bruylants, T., Stijns, E., Ottevaere, H., Schelkens, P., 2013. Wavelet coding of off-axis holographic images. Proc. SPIE 8856, 88561L. http://dx.doi.org/10.1117/12.2027114.

Blinder, D., Bruylants, T., Ottevaere, H., Munteanu, A., Schelkens, P., 2014. JPEG 2000-based compression of fringe patterns for digital holographic microscopy. Opt. Eng. 53 (12), 123102. http://dx.doi.org/10.1117/1.OE.53.12.123102.

Blinder, D., Ahar, A., Symeonidou, A., Xing, Y., Bruylants, T., Schreites, C., Pesquet-Popescu, B., Dufaux, F., Munteanu, A., Schelkens, P., 2015. Open access database for experimental validations of holographic compression engines. In: 2015 Seventh International Workshop on Quality of Multimedia Experience (QoMEX), pp. 1-6.

Boulgouris, N., Strintzis, M., 2002. A family of wavelet-based stereo image coders. IEEE Trans. Circuits Syst. Video Technol. 12 (10), 898-903.

Bove, V., 2012. Display holography's digital second act. Proc. IEEE 100 (4), 918-928. http://dx.doi.org/10.1109/JPROC.2011.2182071.

Bove Jr., V., Plesniak, W., Quentmeyer, T., Barabas, J., 2005. Real-time holographic video images with commodity PC hardware. In: SPIE—The International Society of Optical Engineering.

Bräuer, R., Wyrowski, F., Bryngdahl, O., 1991. Diffusers in digital holography. J. Opt. Soc. Am. A 8 (3), 572-578. http://dx.doi.org/10.1364/JOSAA.8.000572. URL http://josaa.osa.org/abstract.cfm?URI=josaa-8-3-572.

Chang, C.L., Girod, B., 2007. Direction-adaptive discrete wavelet transform for image compression. IEEE Trans. Image Process. 16 (5), 1289-1302.

Chappelier, V., Guillemot, C., 2006. Oriented wavelet transform for image compression and denoising. IEEE Trans. Image Process. 15 (10), 2892-2903. http://dx.doi.org/10.1109/TIP.2006.877526.

Chen, R.H.Y., Wilkinson, T.D., 2009. Computer generated hologram from point cloud using graphics processor. Appl. Opt. 48 (36), 6841-6850. http://dx.doi.org/10.1364/AO.48.006841. URL http://ao.osa.org/abstract.cfm?URI=ao-48-36-6841.

Dallas, W.J., 1980. Computer-generated holograms. In: Frieden, B. (Ed.), The Computer in Optical Research, Topics in Applied Physics, vol. 41. Springer, Berlin, Heidelberg, pp. 291-366, URL http://dx.doi.org/10.1007/BFb0040187.

Darakis, E., Naughton, T.J., 2009. Compression of digital hologram sequences using MPEG-4. In: Proc. SPIE, Holography: Advances and Modern Trends, vol. 7358, p. 735811, URL http://dx.doi.org/10.1117/12.820632.

Darakis, E., Soraghan, J., 2006a. Use of Fresnelets for phase-shifting digital hologram compression. IEEE Trans. Image Process. 15 (12), 3804-3811. http://dx.doi.org/10.1109/TIP.2006.884918.

Darakis, E., Soraghan, J.J., 2006b. Compression of interference patterns with application to phase-shifting digital holography. Appl. Opt. 45 (11), 2437-2443. http://dx.doi.org/10.1364/AO.45.002437. URL http://ao.osa.org/abstract.cfm?URI=ao-45-11-2437.

Darakis, E., Naughton, T.J., Soraghan, J.J., Javidi, B., 2006. Measurement of compression defects in phase-shifting digital holographic data. In: Proc. SPIE, Optical Information Systems IV, vol. 6311, URL http://dx.doi.org/10.1117/12.679445.

Darakis, E., Kowiel, M., Näsänen, R., Naughton, T.J., 2010. Visually lossless compression of digital hologram sequences. In: Proc. SPIE, Image Quality and System Performance VII.

Daubechies, I., 1992. Ten Lectures on Wavelets. Society for Industrial and Applied Mathematics. http://dx.doi.org/10.1137/1.9781611970104. URL http://epubs.siam.org/doi/abs/10.1137/1.9781611970104.

Dirac, 2009. Dirac video compression. URL http://www.diracvideo.org/.

Dufaux, F., Pesquet-Popescu, B., Cagnazzo, M., 2013. John Wiley & Sons, Ltd. ISBN 9781118583593. http://dx.doi.org/10.1002/9781118583593.fmatter.

Finke, G., Kujawińska, M., Kozacki, T., 2015. Visual perception in multi SLM holographic displays. Appl. Opt. 54 (12), 3560-3568. http://dx.doi.org/10.1364/AO.54.003560. URL http://ao.osa.org/abstract.cfm?URI=ao-54-12-3560.

Gabor, D., 1948. A new microscopic principle. Nature 161 (4098), 777-778.

Goodman, J., 1996. Introduction to Fourier Optics, second ed. McGraw-Hill, New York.

Goodman, J., Lawrence, R., 1967. Digital image formation from electronically detected holograms. Appl. Phys. Lett. 11 (3), 77-79. http://dx.doi.org/10.1063/1.1755043. URL http://scitation.aip.org/content/aip/journal/apl/11/3/10.1063/1.1755043.

Gouze, A., Antonini, M., Barlaud, M., Macq, B., 2004. Design of signal-adapted multidimensional lifting scheme for lossy coding. IEEE Trans. Image Process. 13 (12), 1589-1603. http://dx.doi.org/10.1109/TIP.2004.837556.

Hampson, F.J., Pesquet, J.C., 1998. M-band nonlinear subband decompositions with perfect reconstruction. IEEE Trans. Image Process. 7 (11), 1547-1560.

Ito, T., Eldeib, H., Yoshida, K., Takahashi, S., Yabe, T., Kunugi, T., 1996. Special-purpose computer for holography HORN-2. Comput. Phys. Commun. 93 (1), 13-20. http://dx.doi.org/10.1016/0010-4655(95)00125-5. URL http://www.sciencedirect.com/science/article/pii/0010465595001255.

Ito, T., Masuda, N., Yoshimura, K., Shiraki, A., Shimobaba, T., Sugie, T., 2005. Special-purpose computer HORN-5 for a real-time electroholography. Opt. Express 13 (6), 1923-1932. http://dx.doi.org/10.1364/OPEX.13.001923. URL http://www.opticsexpress.org/abstract.cfm?URI=oe-13-6-1923.

Kaaniche, M., Benazza-Benyahia, A., Pesquet-Popescu, B., Pesquet, J.C., 2009. Vector lifting schemes for stereo image coding. IEEE Trans. Image Process. 18 (11), 2463-2475.

Kaaniche, M., Benazza-Benyahia, A., Pesquet-Popescu, B., Pesquet, J.C., 2011a. Non separable lifting scheme with adaptive update step for still and stereo image coding. Signal Process. 91 (12), 2767-2782.

Kaaniche, M., Benazza-Benyahia, A., Pesquet-Popescu, B., Pesquet, J.C., 2011b. Non-separable lifting scheme with adaptive update step for still and stereo image coding. Signal Process. 91 (12), 2767-2782. doi:http://dx.doi.org/10.1016/j.sigpro.2011.01.003. Advances in Multirate Filter Bank Structures and Multiscale Representations, URL http://www.sciencedirect.com/science/article/pii/S0165168411000041.

Kaaniche, M., Pesquet-Popescu, B., Benazza-Benyahia, A., Pesquet, J.C., 2012. Adaptive lifting scheme with sparse criteria for image coding. J. Adv. Signal Process. 2012 (1), 10.

Kim, H., Hahn, J., Lee, B., 2008. Mathematical modeling of triangle-mesh-modeled three-dimensional surface objects for digital holography. Appl. Opt. 47 (19), D117-D127. http://dx.doi.org/10.1364/AO.47.00D117. URL http://ao.osa.org/abstract.cfm?URI=ao-47-19-D117.

Lee, S.K., Hong, S.I., Kim, Y.S., Lim, H.G., Jo, N.Y., Park, J.H., 2013. Hologram synthesis of three-dimensional real objects using portable integral imaging camera. Opt. Express 21 (20), 23662-23670. http://dx.doi.org/10.1364/OE.21.023662. URL http://www.opticsexpress.org/abstract.cfm?URI=oe-21-20-23662.

Leith, E.N., Upatnieks, J., 1962. Reconstructed wavefronts and communication theory. J. Opt. Soc. Am. 52 (10), 1123-1128. http://dx.doi.org/10.1364/JOSA.52.001123. URL http://www.opticsinfobase.org/abstract.cfm?URI=josa-52-10-1123.

Liebling, M., Blu, T., Unser, M., 2003. Fresnelets: new multiresolution wavelet bases for digital holography. IEEE Trans. Image Process. 12 (1), 29-43. http://dx.doi.org/10.1109/TIP.2002.806243.

Linde, Y., Buzo, A., Gray, R., 1980. An algorithm for vector quantizer design. IEEE Trans. Commun. 28 (1), 84-95. http://dx.doi.org/10.1109/TCOM.1980.1094577.

Liu, J.P., 2012. Controlling the aliasing by zero-padding in the digital calculation of the scalar diffraction. J. Opt. Soc. Am. A 29 (9), 1956-1964. http://dx.doi.org/10.1364/JOSAA.29.001956. URL http://josaa.osa.org/abstract.cfm?URI=josaa-29-9-1956.

Lloyd, S., 1982. Least squares quantization in PCM. IEEE Trans. Inf. Theor. 28 (2), 129-137. http://dx.doi.org/10.1109/TIT.1982.1056489.

Lucente, M., 1993. Interactive computation of holograms using a look-up table. J. Electron. Imaging 2, 28-34.

Mallat, S., 2009. Geometrical grouplets. Appl. Comput. Harmonic Anal. 26 (2), 161-180. doi:http://dx.doi.org/10.1016/j.acha.2008.03.004. URL http://www.sciencedirect.com/science/article/pii/S1063520308000444.

Masuda, N., Ito, T., Tanaka, T., Shiraki, A., Sugie, T., 2006. Computer generated holography using a graphics processing unit. Opt. Express 14 (2), 603-608. http://dx.doi.org/10.1364/OPEX.14.000603. URL http://www.opticsexpress.org/abstract.cfm?URI=oe-14-2-603.

Matsushima, K., 2005. Computer-generated holograms for three-dimensional surface objects with shade and texture. Appl. Opt. 44 (22), 4607-4614. http://dx.doi.org/10.1364/AO.44.004607. URL http://ao.osa.org/abstract.cfm?URI=ao-44-22-4607.

Matsushima, K., Kondoh, A., 2004. A wave-optical algorithm for hidden-surface removal in digitally synthetic full-parallax holograms for three-dimensional objects. Proc. SPIE 5290, 90-97. http://dx.doi.org/10.1117/12.526747.

Matsushima, K., Nakahara, S., 2009. Extremely high-definition full-parallax computer-generated hologram created by the polygon-based method. Appl. Opt. 48 (34), H54-H63. http://dx.doi.org/10.1364/AO.48.000H54. URL http://ao.osa.org/abstract.cfm?URI=ao-48-34-H54.

Max, J., 1960. Quantizing for minimum distortion. IRE Trans. Inf. Theory 6 (1), 7-12. http://dx.doi.org/10.1109/TIT.1960.1057548.

Mills, G., Yamaguchi, I., 2005. Effects of quantization in phase-shifting digital holography. Appl. Opt. 44 (7), 1216-1225. http://dx.doi.org/10.1364/AO.44.001216. URL http://ao.osa.org/abstract.cfm?URI=ao-44-7-1216.

Mishina, T., Okui, M., Okano, F., 2006. Calculation of holograms from elemental images captured by integral photography. Appl. Opt. 45 (17), 4026-4036. http://dx.doi.org/10.1364/AO.45.004026. URL http://ao.osa.org/abstract.cfm?URI=ao-45-17-4026.

Moellenhoff, M., Maier, M., 1998. Transform coding of stereo image residuals. IEEE Trans. Image Process. 7 (6), 804-812. http://dx.doi.org/10.1109/83.679421.

Naughton, T., Frauel, Y., Javidi, B., Tajahuerce, E., 2002. Compression of digital holograms for three-dimensional object reconstruction and recognition. Appl. Opt. 41 (20), 4124-4132.

Naughton, T.J., McDonald, J.B., Javidi, B., 2003. Efficient compression of Fresnel fields for internet transmission of three-dimensional images. Appl. Opt. 42 (23), 4758-4764. http://dx.doi.org/10.1364/AO.42.004758. URL http://ao.osa.org/abstract.cfm?URI=ao-42-23-4758.

Onural, L., Ozaktas, H.M., 2007. Signal processing issues in diffraction and holographic 3DTV. Image Commun. 22 (2), 169-177. http://dx.doi.org/10.1016/j.image.2006.11.010.

Onural, L., Gotchev, A.P., Ozaktas, H.M., Stoykova, E., 2007. A survey of signal processing problems and tools in holographic three-dimensional television. IEEE Trans. Circuits Syst. Video Technol. 17 (11), 1631-1647.

Park, J.H., Kim, M.S., Baasantseren, G., Kim, N., 2009. Fresnel and Fourier hologram generation using orthographic projection images. Opt. Express 17 (8), 6320-6334. http://dx.doi.org/10.1364/OE.17.006320. URL http://www.opticsexpress.org/abstract.cfm?URI=oe-17-8-6320.

Pereira, F., Ebrahimi, T., 2002. The MPEG-4 Book. Prentice Hall, Upper Saddle River, NJ.

Ritter, A., Böttger, J., Deussen, O., König, M., Strothotte, T., 1999. Hardware-based rendering of full-parallax synthetic holograms. Appl. Opt. 38, 1364-1369.

Said, A., Pearlman, W., 1996. A new, fast, and efficient image codec based on set partitioning in hierarchical trees. IEEE Trans. Circuits Syst. Video Technol. 6 (3), 243-250. http://dx.doi.org/10.1109/76.499834.

Sando, Y., Itoh, M., Yatagai, T., 2003. Holographic three-dimensional display synthesized from three-dimensional Fourier spectra of real existing objects. Opt. Lett. 28 (24), 2518-2520. http://dx.doi.org/10.1364/OL.28.002518. URL http://ol.osa.org/abstract.cfm?URI=ol-28-24-2518.

Schelkens, P., Skodras, A., Ebrahimi, T., 2009. The JPEG 2000 Suite. Wiley, Chichester.

Schnars, U., Jüptner, W., 1994. Direct recording of holograms by a CCD target and numerical reconstruction. Appl. Opt. 33 (2), 179-181. http://dx.doi.org/10.1364/AO.33.000179. URL http://ao.osa.org/abstract.cfm?URI=ao-33-2-179.

Senoh, T., Wakunami, K., Ichihashi, Y., Sasaki, H., Oi, R., Yamamoto, K., 2014. Multi-view image and depth map coding for holographic TV system. Opt. Eng. 53 (11), 112302. http://dx.doi.org/10.1117/1.OE.53.11.112302.

Seo, Y.H., Choi, H.J., Bae, J.W., Yoo, J.S., Kim, D.W., 2006. Data compression technique for digital holograms using a temporally scalable coding method for 2-D images. In: 2006 IEEE International Symposium on Signal Processing and Information Technology, pp. 326-331.

Seo, Y.H., Choi, H.J., Kim, D.W., 2007. 3D scanning-based compression technique for digital hologram video. Image Commun. 22 (2), 144-156. http://dx.doi.org/10.1016/j.image.2006.11.007.

Shaked, N.T., Rosen, J., 2008. Modified Fresnel computer-generated hologram directly recorded by multiple-viewpoint projections. Appl. Opt. 47 (19), D21-D27. http://dx.doi.org/10.1364/AO.47.000D21. URL http://ao.osa.org/abstract.cfm?URI=ao-47-19-D21.

Shaked, N.T., Katz, B., Rosen, J., 2009. Review of three-dimensional holographic imaging by multiple-viewpoint-projection based methods. Appl. Opt. 48 (34), H120-H136. http://dx.doi.org/10.1364/AO.48.00H120. URL http://ao.osa.org/abstract.cfm?URI=ao-48-34-H120.

Shapiro, J., 1993. Embedded image coding using zerotrees of wavelet coefficients. IEEE Trans. Signal Process. 41 (12), 3445-3462. http://dx.doi.org/10.1109/78.258085.

Shimobaba, T., Masuda, N., Sugie, T., Hosono, S., Tsukui, S., Ito, T., 2000. Special-purpose computer for holography HORN-3 with PLD technology. Comput. Phys. Commun. 130, 75-82. doi:http://dx.doi.org/10.1016/S0010-4655(00)00044-8. URL http://www.sciencedirect.com/science/article/pii/S0010465500000448.

Shimobaba, T., Hishinuma, S., Ito, T., 2002. Special-purpose computer for holography HORN-4 with recurrence algorithm. Comput. Phys. Commun. 148 (2), 160-170. doi:http://dx.doi.org/10.1016/S0010-4655(02)00473-3. URL http://www.sciencedirect.com/science/article/pii/S0010465502004733.

Shortt, A., Naughton, T.J., Javidi, B., 2006a. Compression of digital holograms of three-dimensional objects using wavelets. Opt. Express 14 (7), 2625-2630. http://dx.doi.org/10.1364/OE.14.002625. URL http://www.opticsexpress.org/abstract.cfm?URI=oe-14-7-2625.

Shortt, A.E., Naughton, T.J., Javidi, B., 2006b. A companding approach for nonuniform quantization of digital holograms of three-dimensional objects. Opt. Express 14 (12), 5129-5134. http://dx.doi.org/10.1364/OE.14.005129. URL http://www.opticsexpress.org/abstract.cfm?URI=oe-14-12-5129.

Shortt, A., Naughton, T., Javidi, B., 2007. Histogram approaches for lossy compression of digital holograms of three-dimensional objects. IEEE Trans. Image Process. 16 (6), 1548-1556. http://dx.doi.org/10.1109/TIP.2007.894269.

Stein, A.D., Wang, Z., Leigh, J.S., 1992. Computer-generated holograms: a simplified ray-tracing approach. In: Computers in Physics, pp. 389-392.

Stevens, A.R., Munteanu, J.C., Schelkens, P., 2008. Optimized directional lifting with reduced complexity. In: EURASIP European Signal Processing Conference, EUSIPCO 2008, Lausanne, Switzerland, p. 5.

Sullivan, G., Ohm, J., Han, W.J., Wiegand, T., 2012. Overview of the high efficiency video coding (HEVC) standard. IEEE Trans. Circuits Syst. Video Technol. 22 (12), 1649-1668. http://dx.doi.org/10.1109/TCSVT.2012.2221191.

Sweldens, W., 1996. The lifting scheme: a custom-design construction of biorthogonal wavelets. Appl. Comput. Harmonic Anal. 3 (2), 186-200.

Taubman, D., 2000. High performance scalable image compression with EBCOT. IEEE Trans. Image Process. 9 (7), 1158-1170. http://dx.doi.org/10.1109/83.847830.

Taubman, D., Marcellin, M., 2001. JPEG2000: Image Compression Fundamentals, Standards and Practice. Kluwer Academic Publishers, Norwell, MA, USA.

Tricoles, G., 1987. Computer generated holograms: an historical review. Appl. Opt. 26 (20), 4351-4360. http://dx.doi.org/10.1364/AO.26.004351. URL http://ao.osa.org/abstract.cfm?URI=ao-26-20-4351.

Unser, M., 1999. Splines: a perfect fit for signal and image processing. IEEE Signal Process. Mag. 16 (6), 22-38. http://dx.doi.org/10.1109/79.799930.

Viswanathan, K., Gioia, P., Morin, L., 2013. Wavelet compression of digital holograms: towards a view-dependent framework. Proc. SPIE 8856. http://dx.doi.org/10.1117/12.2027199.

Viswanathan, K., Gioia, P., Morin, L., 2014. Morlet wavelet transformed holograms for numerical adaptive view-based reconstruction. Proc. SPIE 9216, 92160G. http://dx.doi.org/10.1117/12.2061588.

Wakunami, K., Yamaguchi, M., 2011. Calculation for computer generated hologram using ray-sampling plane. Opt. Express 19 (10), 9086-9101. http://dx.doi.org/10.1364/OE.19.009086. URL http://www.opticsexpress.org/abstract.cfm?URI=oe-19-10-9086.

Wiegand, T., Sullivan, G., Bjontegaard, G., Luthra, A., 2003. Overview of the H.264/AVC video coding standard. IEEE Trans. Circuits Syst. Video Technol. 13 (7), 560-576. http://dx.doi.org/10.1109/TCSVT.2003.815165.

Xing, Y., Pesquet-Popescu, B., Dufaux, F., 2013a. Compression of computer generated hologram based on phase-shifting algorithm. In: European Workshop on Visual Information Processing, pp. 172-177.

Xing, Y., Pesquet-Popescu, B., Dufaux, F., 2013b. Compression of computer generated phase-shifting hologram sequence using AVC and HEVC. In: Proc. SPIE, Applications of Digital Image Processing XXXVI, vol. 8856, San Diego, CA, USA.

Xing, Y., Kaaniche, M., Pesquet-Popescu, B., Dufaux, F., 2014a. Vector lifting scheme for phase-shifting holographic data compression. Opt. Eng. 53 (11), 112312. http://dx.doi.org/10.1117/1.OE.53.11.112312.

Xing, Y., Pesquet-Popescu, B., Dufaux, F., 2014b. Comparative study of scalar and vector quantization on different phase-shifting digital holographic data representations. In: 3DTV-Conference: The True Vision—Capture, Transmission and Display of 3D Video (3DTV-CON), 2014, pp. 1-4.

Xing, Y., Kaaniche, M., Pesquet-Popescu, B., Dufaux, F., 2015. Adaptive nonseparable vector lifting scheme for digital holographic data compression. Appl. Opt. 54 (1), A98-A109. http://dx.doi.org/10.1364/AO.54.000A98. URL http://ao.osa.org/abstract.cfm?URI=ao-54-1-A98.

Yamaguchi, M., 2011. Ray-based and wavefront-based holographic displays for high-density light-field reproduction. Proc. SPIE 8043, 804306. http://dx.doi.org/10.1117/12.884500.

Yamaguchi, I., Zhang, T., 1997. Phase-shifting digital holography. Opt. Lett. 22 (16), 1268-1270. http://dx.doi.org/10.1364/OL.22.001268. URL http://ol.osa.org/abstract.cfm?URI=ol-22-16-1268.

Zhong, J., Weng, J., Hu, C., 2009. Reconstruction of digital hologram by use of the wavelet transform. In: Advances in Imaging. Optical Society of America, p. DWB16, URL http://www.osapublishing.org/abstract.cfm?URI=DH-2009-DWB16.

Printed in the United States
By Bookmasters